Estimating Checklist for Capital Projects

Estimating Checklist for Capital Projects

Second edition

The Association of Cost Engineers

E & FN SPON
An Imprint of Chapman & Hall
London · New York · Tokyo · Melbourne · Madras

UK	Chapman & Hall, 2–6 Boundary Row, London SE1 8HN
USA	Van Nostrand Reinhold, 115 5th Avenue, New York, NY 10003
JAPAN	Chapman & Hall Japan, Thomson Publishing Japan, Hirakawacho Nemoto Building, 7F, 1-7-11 Hirakawa-cho, Chiyoda-ku, Tokyo 102
AUSTRALIA	Chapman & Hall Australia, Thomas Nelson Australia, 102 Dodds Street, South Melbourne, Victoria 3205
INDIA	Chapman & Hall India, R. Seshadri, 32 Second Main Road, CIT East, Madras 600 035

First edition 1991

© 1991 The Association of Cost Engineers

Typeset in 11/12½ Rockwell by Acorn Bookwork, Salisbury, Wiltshire

Printed in Great Britain by
St Edmundsbury Press Ltd, Bury St Edmunds, Suffolk

ISBN 0 419 15560 0   0 442 30841 8 (USA)

Apart from any fair dealing for the purposes of research or private study, or criticism or review, as permitted under the UK Copyright Designs and Patents Act, 1988, this publication may not be reproduced, stored, or transmitted, in any form or by any means, without the prior permission in writing of the publishers, or in the case of reprographic reproduction only in accordance with the terms of the licences issued by the Copyright Licensing Agency in the UK, or in accordance with the terms of licences issued by the appropriate Reproduction Rights Organization outside the UK. Enquiries concerning reproduction outside the terms stated here should be sent to the publishers at the UK address printed on this page.

The publisher makes no representation, express or implied, with regard to the accuracy of the information contained in this book and cannot accept any legal responsibility or liability for any errors or omissions that may be made.

British Library Cataloguing in Publication Data
Estimating checklist for capital projects – 2nd ed.
   1. Construction industries. Costing
   I. Association of Cost Engineers
   624.0681

ISBN 0–419–15560–0

Library of Congress Cataloging-in-Publication Data
Estimating checklist for capital projects / the Association of Cost
  Engineers. — 1st ed.
    p. cm.
  ISBN 0–442–30841–8
   1. Engineering—Estimates.   I. Association of Cost Engineers.
TA183.E84 1991
620'.0029'9—dc20
                                                                     90–49082
                                                                         CIP

Contents

01.00	Preface	1
02.00	Introduction	3
03.00	Land and site development	5
04.00	Civil engineering	9
05.00	Buildings including building services	19
06.00	Structures	23
07.00	Plant including machinery and equipment	27
08.00	Mechanical services	33
09.00	Electrical services	45
10.00	Instrumentation and controls including communications	53
11.00	Remote fabrication/pre-assembly facilities	65
12.00	Insulation	69
13.00	Protective coatings	71
14.00	Offsite facilities	73
15.00	Import/export pipelines	77
16.00	Import/export loading facilities	83
17.00	Operational and general services	85
18.00	Temporary and/or common site facilities	89
19.00	Infrastructure	93
20.00	Modifications and alterations to existing plant	95
21.00	Spares	97
22.00	Engineering design and procurement	99
23.00	Project construction management and project charges	105
24.00	Contract conditions	109
25.00	Overseas projects	111
26.00	Commissioning, initial operation and training	115
27.00	Provisional sums, contingencies and escalation	117
Appendix	Engineering design and procurement (additional details).	119

Faithful&Gould

Quantity Surveyors

Cost Engineers

Project Managers

Just check us out for yourself

☐ Thirty years unrivalled experience in Heavy Industrial projects of every description

☐ In-depth experience of petrochemical, oil, gas, nuclear, coal and pharmaceutical projects

☐ Staff with a high level of proven expertise and commitment

☐ Comprehensive cost management, from preparing estimates and contracts to dealing with final accounts and settlement of claims

☐ In-house computer department providing bespoke software for Heavy Industrial projects

☐ A network of national, European and international offices

☐ A total quality of service that can't be beaten anywhere

For further details, please contact any partner at one of our offices.

61 Portland Place
London
W1N 3AJ
071 637 2345

Compton House
Westgate
Leeds
LS1 4ND
0532 451535

116 Pembroke Road
Clifton
Bristol
BS8 3EW
0272 738256

4 Pilcher Gate
Nottingham
NG1 1QE
0602 583856

2 Yarm Road
Stockton-on-Tees
TS18 3NA
0642 675136

Westgate House
Hale Road
Altrincham
WA14 2EX
061 941 2831

14 Chester Street
Edinburgh
EH3 7RA
031 226 3132

53 Cathedral Road
Cardiff
CF1 9HD
0222 374059

3 Coed Pella Road
Colwyn Bay
LL29 7AT
0492 534195

50-54 St Paul's Square
Birmingham
B3 1QS
021 236 8040

Hong Kong

Singapore

Toronto

New York

Associated offices

Federal Republic of Germany

Spain

Belgium

Portugal

Switzerland

Brazil

Libya

Nigeria

01.00

Preface

The Association of Cost Engineers promotes the technical study and development of cost engineering and the application of scientific principles and techniques to problems of cost estimating, cost control, profitability and investment appraisal. As part of its objective, the Association published in 1970 an Estimating Checklist for Capital Projects and in 1982 a further Estimating Checklist for Offshore Projects. The present checklist supersedes that for 1970, but does not include offshore projects, for which the 1982 checklist remains available.

As the UK moves towards a more unified European market the Association of Cost Engineers, a founder member of the International Cost Engineering Council, is working closely with European, American and other cost engineering associations to disseminate technical knowledge for international projects.

The aim of this checklist is to provide an *aide-mémoire* for the estimator, cost engineer, quantity surveyor, project control engineer, planner and others responsible for estimating and controlling costs and for ensuring that contracts are completed on time and within budget.

The checklist has been produced by a committee experienced in estimating and cost engineering, and the contents result from the combined efforts of the members with support, guidance and advice from colleagues in their respective companies. The list may not necessarily agree with the views of the individual companies.

The members of the Estimating Checklist Committee were B.G. Wheeler (Chairman), H.S. Mason (Editor), W.L. Bruck, F.W. Connew, J.M. Hart, R.P. Lewington, R. Pickett, R.M. Rear and D. Walters. The co-operation of the following companies in allowing the members of the committee to participate in the work is gratefully acknowledged: BP International Ltd, Bovis Construction Ltd, Brown and Root Vickers Ltd, Currie and Brown International Ltd, John Brown Engineers and Constructors Ltd, Matthew Hall Engineering Ltd, Mobil Services Co. Ltd, National Power and Wimpey Engineering Ltd.

The Association of Cost Engineers, 'Limited by Guarantee', Lea House, 5 Middlewich Road, Sandbach, Cheshire CW11 9XL

Thermal Insulation

Whatever the problem we have the solution...

With operating centres in 5 European countries, Hertel is well established as a leading international contractor to the petrochemical and heavy engineering industries.

Hertel combines a high quality of workmanship, management control and industrial relations together with a comprehensive maintenance service for term maintenance and shut down operation.

The ability to complete projects on time and within budget makes Hertel the natural choice for Thermal insulation contracts.

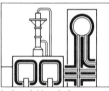
Industrial Insulation

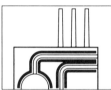
Power Plant

Offshore

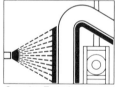
Spraying Techniques

Climate Controlled Room

Acoustic Insulation

United Kingdom:-
Hertel (U.K.) Ltd., Wallis Road,
Skippers Lane, Middlesborough,
Cleveland TS6 6JB.
Tel: 0642 467652/469532 Fax: 0642 452911
Centres: St. Helens, Goole, Hereford, Bridgend Mon.

also: Belgium, France, Germany, Netherlands.

02.00

Introduction

This checklist is fundamentally a list of work items, although reference is also made to Conditions of Contract, insurances, etc. It does not contain a list of pricing considerations, and when calculating the correct price level, due allowance must be made for such items as labour and material availability, pay levels, productivity, market conditions, site agreements, working restrictions, climate, escalation, risk and specifications relating to the particular project.

Various sections of the construction industry use different terms and practices in formulating contracts. On a building project the work around the building is usually included as an external works section, but for an oil, gas or chemical plant the work beyond the plot limits (or battery limits) of the main process plant is normally called the offsite facilities or offsites. The offsite facilities typically include all roads, pipework, cabling and services to and from the plot limits, preliminary treatment facilities, sewage treatment, firewater, flare, storage tanks, administration buildings, substations and other facilities required to support the process plant. In other industries the external works or offsite facilities are included in the appropriate work section. Thus roads and paths may appear in civil engineering, external works for buildings, or offsite facilities. Plant and equipment may similarly be included in plant, mechanical services, offsite facilities, or import/export loading facilities. Some duplication of items within the check will thus occur and the item may not always be within the section familiar to the user.

Construction contracts are not standard and the differences arising from the type of construction, building, plant or process, together with special requirements relating to climate, local regulations, conditions, customs and resources, are numerous. This checklist should therefore not be considered as a complete list of every item that may occur, and is no substitute for ascertaining the exact requirements and particular peculiarities of any contract, but should be used simply as an *aide-mémoire* and further check against the list prepared by the estimator.

The Contents list shows the book structure and is such that each trade appears as a separate section in an order similar to that usually found in Specifications and Bills of Quantities and closely related to the order of construction. At the end of the book there are sections for Special Facilities, Engineering Design and Sundry Items which relate to all trades or to special considerations for working abroad.

Reference should always be made back to the main trades when dealing with a multi-trade section as for example it is not practical to repeat civils and pipework in full in all the sections in which they occur.

A digit reference system is used starting with the Section number.

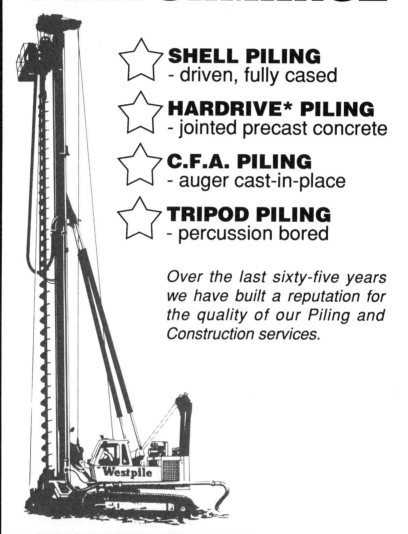

03.00

Land and site development

03.01 Land acquisition
03.02 Rates, taxes, stamp duty, levies, tariffs
03.03 Lump sum contributions to local authorities
03.04 Compensation for damage
03.05 Fees (e.g. legal)
03.06 Access to site including wayleaves and easements
03.07 Services to and from site including wayleaves and easements
03.08 Surveys and investigations
03.09 Demolition, site clearance and diversions
03.10 Ground stabilization
03.11 Fencing
03.12 Signs and illuminated boards
03.13 Site screening
03.14 Pre-production mine development

This section concerns those items connected with preparation of the proposed site before the main construction. Activities both above ground and underground are considered. Some items overlap into section 04.00 Civil engineering.

03.01 LAND ACQUISITION

03.01.01 Purchase
03.01.02 Freehold
03.01.03 Leasehold
03.01.04 Rental
03.01.05 Development grants or tax concessions obtainable

03.02 RATES, TAXES, STAMP DUTY, LEVIES, TARIFFS

03.03 LUMP SUM CONTRIBUTIONS TO LOCAL AUTHORITIES

03.04 COMPENSATION FOR DAMAGE

03.05 FEES (e.g. LEGAL)

03.06 ACCESS TO SITE INCLUDING WAYLEAVES AND EASEMENTS

03.06.01 Roads
03.06.02 Footpaths
03.06.03 Rail tracks
03.06.04 Conveyors
03.06.05 Canals and waterways
03.06.06 Landing strips and helicopter pads
03.06.07 Tunnels and bridges

03.07 SERVICES TO AND FROM SITE INCLUDING WAYLEAVES AND EASEMENTS

03.07.01 Electricity
03.07.02 Gas
03.07.03 Water
03.07.04 Telephone and communications
03.07.05 Sewer
03.07.06 Pipelines (see also section 15.00)

03.08 SURVEYS AND INVESTIGATIONS

03.08.01 Ecological and environmental surveys
03.08.02 Geological, geotechnical, hydrographic and seismic surveys
03.08.03 Trial holes, boreholes, samples, analyses and tests
03.08.04 Site instrumental tests

03.09 DEMOLITION, SITE CLEARANCE AND DIVERSIONS

03.09.01 Demolition and site clearance (see also section 04.00)
03.09.02 Diversion of roads, pipelines and services
03.09.03 Diversion of canals and waterways
03.09.04 Relocation of plant/buildings/townships etc.
03.09.05 Protection of adjoining properties
03.09.06 Removal of underground obstructions

03.10 GROUND STABILIZATION

03.10.01 Ground protection, compaction and/or grouting
03.10.02 Ground stabilization
03.10.03 Sink holes, subsidence and fault lines
03.10.04 Diaphragm walls
03.10.05 Suppression of vegetation
03.10.06 Preliminary drainage, dewatering or sand drains

03.11 FENCING

03.11.01 Permanent and temporary fencing (see also section 04.00)
03.11.02 Gates (see also section 04.00)
03.11.03 Hoardings

03.12 SIGNS AND ILLUMINATED BOARDS

| 03.13 | SITE SCREENING |

03.13.01 Screening embankments (see also section 04.00)
03.13.02 Tree/bush planting (see also section 04.00)

| 03.14 | PRE-PRODUCTION MINE DEVELOPMENT |

03.14.01 Shafts/bores/raises/adits
03.14.02 Drives/gates/headings
03.14.03 Rooms/stopes

Lilley Construction

Civil Engineering and Building Contractors to the Oil, Gas & Petrochemical Industries.

Major projects undertaken at:-

Barrow-in-Furness, Dimlington, Fawley, Flotta, Grangemouth, Hamble, Milford Haven, Mossmorran, St Fergus.

Lilley Construction Limited, 331 Charles Street, Glasgow G21 2QX
Tel 041 - 552 - 6565
A MEMBER OF THE LILLEY GROUP

LILLEY

04.00

Civil engineering

04.01 Excavation
04.02 Filling and shoring
04.03 Geotechniques
04.04 Concrete supply, yard facilities, formwork and sundries
04.05 Foundations
04.06 Pits, basins and channels
04.07 Ducts etc.
04.08 Steel encasement
04.09 Transport facilities and paving
04.10 Marine work
04.11 Structures
04.12 Cooling towers
04.13 Mining
04.14 Drainage
04.15 Water supply
04.16 Pipework and cabling
04.17 Brickwork, blockwork and masonry
04.18 Landscaping
04.19 Roads

See also section 05.00 Buildings and section 14.02 Civil engineering work in connection with offsite facilities.

04.01 EXCAVATION

04.01.01 Site clearance, demolition
04.01.02 Excavation of topsoil including lifting turf
04.01.03 Reduce level excavation or cut and fill
04.01.04 Excavation of soft spots and filling
04.01.05 Rock removal including ripping, drilling and blasting
04.01.06 Excavation for foundations and basements
04.01.07 Excavation for existing services
04.01.08 Excavation of pits, basins, channels and ducts
04.01.09 Excavation for transport facilities and paving
04.01.10 Excavation for exposed pipe or cable tracks
04.01.11 Excavation for pipes

04.01.12 Excavation for cables
04.01.13 Diaphragm walls (see also section 03.00)
04.01.14 Dewatering (see also section 03.00)
04.01.15 Forming earth bunds and embankments
04.01.16 Sand padding (if not included elsewhere)
04.01.17 Disposal of surplus including offsite tips
04.01.18 Double handling of surplus

04.02 FILLING AND SHORING

04.02.01 Selected, non-selected or imported filling
04.02.02 Bulk fill: embankments, terraces, earth bunds
04.02.03 Under structures, pavings and slabs
04.02.04 Make up levels and soft spots
04.02.05 Borrow pits and quarries
04.02.06 Blinding
04.02.07 Compaction
04.02.08 Sand padding to pipes and cables
04.02.09 Temporary shoring or support work
      (a) timber
      (b) steel
      (c) scaffolding

04.03 GEOTECHNIQUES

04.03.01 Piling
      (a) bored, cast or driven
      (b) precast concrete, prestressed and precast planks
      (c) timber
      (d) steel-isolated and interlocking sheet
      (e) cofferdams and caissons
      (f) pre-boring and disposal of surplus
      (g) enlarged bases
      (h) jetting
      (j) obstructions
      (k) permanent casings
      (l) concrete or backfill to empty bore

- (m) trim heads
- (n) excavation, concrete, reinforcement and formwork
- (p) tests: Kentledge, non-destructive, sacrificial provisions; penetration, compression, lateral, uplift; inclinometers
- (q) delays: rigs standing

04.03.02 Diaphragm walls
- (a) excavation, disposal, backfill, concrete, reinforcement
- (b) trim level of wall
- (c) joints
- (d) guide walls: excavation, disposal, backfill, concrete, reinforcement and formwork
- (e) tests
- (f) delays: rigs standing

04.03.03 Underpinning
- (a) excavation, disposal, support work, breaking out existing, backfilling concrete, formwork, reinforcement, brickwork, tanking

04.03.04 Ground and rock anchors
- (a) drill grout holes, drive injection pipes, grouting

04.03.05 Gun-applied concrete
- (a) slabs, walls, beams, columns, sprayed concrete, mesh reinforcement and expanded metal

04.03.06 Reinforced earth
- (a) strips, planks, grids, anchors, fabrics, GRP, weldmesh, expanded metal

04.04 CONCRETE SUPPLY, YARD FACILITIES, FORMWORK AND SUNDRIES

04.04.01 Concrete supply: sitemix, premix, pumped, additives: waterproof, sulphate resisting

04.04.02 Site set up: batching plant, hardstandings, aggregate bins, cement silo, debagging machines, water supply, storage tanks

04.04.03 Transport: mixer trucks, dump trucks, dumpers, conveyors, monorail, pumps, chutes

04.04.04 Vibrators: vibrators, pokers, screed boards

04.04.05 Cranage
- (a) tower, fixed, rail mounted, climbing
- (b) mobile, wheel based, tracked
- (c) fork-lift trucks

04.04.06 Formwork/shuttering
- (a) timber faced, steel panels
- (b) face finish: sawn, fair, other
- (c) decking: horizontal, sloping, domed, vertical, battered, circular, coffered, troughed, voids, woodwool, metal profiled
- (d) vertical: beams, columns, walls

04.04.07 Reinforcement
- (a) mild steel, high yield steel, stainless steel
- (b) fabric: laps and joints
- (c) prestressed: pre-tensioned/post-tensioned

04.04.08 Precast concrete
- (a) beams, columns, slabs, portal frames
- (b) segmental units: subways, culverts, ducts
- (c) copings, sills, cappings, lintels
- (d) kerbs, pavings, channels, edgings
- (e) reinforced: prestressed: pre-tensioned, post-tensioned

04.04.09 Sundries
- (a) expansion and contraction joints
- (b) waterstops
- (c) holding-down bolts
- (d) waterproof tanking
- (e) surface treatments and finishes
- (f) chemical or acid-resistant tiling
- (g) cutting, chasing, rebating
- (h) geotextile fabrics and sheetings
- (i) grouting

04.05 FOUNDATIONS

04.05.01 Foundations for structures
04.05.02 Retaining and basement walls
04.05.03 Foundations, paving and bunds to and around tanks
04.05.04 Precast products, including beams and sleepers
04.05.05 Foundations for plant, machinery, equipment and pipework
04.05.06 Foundations for electrical and instrumentation installations
04.05.07 Foundations for import/export loading facilities

04.06 PITS, BASINS AND CHANNELS

04.06.01 Firewater
04.06.02 Sewage treatment
04.06.03 Oily water separators
04.06.04 Slurry ponds
04.06.05 Weighbridge pits
04.06.06 Cooling water including intake/outfall structures, culverts, strainer, syphon recovery and make-up pump chambers
04.06.07 Rail unloader pits

04.07 DUCTS ETC.

04.07.01 Cooling water culverts (if not in Section 04.10)
04.07.02 Cable ducts
04.07.03 Pipe ducts
04.07.04 Pipe culverts (under roads etc.)
04.07.05 Precast concrete cable tiles and marker posts

04.08 STEEL ENCASEMENT

04.08.01 Steel encasement
04.08.02 Fireproofing

04.09 TRANSPORT FACILITIES AND PAVING

04.09.01 Roads, turning circles and ramps
04.09.02 Paths
04.09.03 Car parks, paved areas and hardstandings
04.09.04 Curbs
04.09.05 Paving under plant
04.09.06 Rail track, platforms and ancillaries (see also section 16.00)
04.09.07 Road/rail canal unloading facilities (see also section 16.00)
04.09.08 Bridges and tunnels
04.09.09 Airfield landing strips and helipads
04.09.10 Traffic signs and markings

04.10 MARINE WORK

- 04.10.01 Cooling water intakes/outfalls, structures and culverts (if not in Section 04.06.07)
- 04.10.02 Tidal work
- 04.10.03 Dredging
- 04.10.04 Hardfilling
- 04.10.05 Jetties and moorings
- 04.10.06 Wharves
- 04.10.07 Reservoirs
- 04.10.08 Dams
- 04.10.09 Breakwaters
- 04.10.10 Foreshore protection
- 04.10.11 Pontoons

04.11 STRUCTURES

- 04.11.01 Towers and silos
- 04.11.02 Windshields, stacks and flues
- 04.11.03 Tanks and pits
- 04.11.04 Bunds and firewalls
- 04.11.05 Storage for raw material or product
- 04.11.06 Loading/unloading structures including lifts, gantries, booms and special structures (see also section 16.00)

04.12 COOLING TOWERS

- 04.12.01 Foundations
- 04.12.02 Shell
- 04.12.03 Collection surround
- 04.12.04 Internal discharge structures
- 04.12.05 Packing
- 04.12.06 Foundations for forced-draught fans

04.13 MINING

04.13.01 Shafts, bores and raises
04.13.02 Tunnels, rooms and stopes
04.13.03 Service tunnels
04.13.04 Storage caverns
04.13.05 Adits
04.13.06 Drives, gates and headings

04.14 DRAINAGE

04.14.01 Stream or channel diversions
04.14.02 Land drainage and open ditches, lined or unlined
04.14.03 Surface water drainage
04.14.04 Sewer drainage
04.14.05 Oily water drainage
04.14.06 Other drainage
04.14.07 Crossings including river, road, rail, fence open cut or thrust
04.14.08 Pipe including fittings and coatings
04.14.09 Paddings, beds and surrounds
04.14.10 Concrete stools, thrust blocks and supports
04.14.11 Manholes, inspection chambers and gullies
04.14.12 Pits, basins and channels (see also section 04.06)
04.14.13 Septic tanks, soakaways and cesspits
04.14.14 Filter beds
04.14.15 Sedimentation chambers
04.14.16 Pumping stations
04.14.17 Valves
04.14.18 Outfalls

04.15 WATER SUPPLY

04.15.01 Dams
04.15.02 Reservoirs
04.15.03 Stream or channel diversions
04.15.04 Leats
04.15.05 Pipework
04.15.06 Manholes and inspection chambers

04.15.07 Compensation water channels
04.15.08 Settling tanks
04.15.09 Sand filters
04.15.10 Chlorination and chemical treatment
04.15.11 Pumping stations
04.15.12 Valves
04.15.13 Intakes

04.16   PIPEWORK AND CABLING *(SEE ALSO SECTIONS 08.00, 09.00 and 10.00)*

04.16.01 Water
04.16.02 Firemains and hydrants
04.16.03 Process or any other pipework below ground (see also section 08.00)
04.16.04 Valves and pits
04.16.05 Pipe coatings
04.16.06 Pipe surrounds
04.16.07 Cable trenches, ducts, padding, tiles and markers

04.17   BRICKWORK, BLOCKWORK AND MASONRY

04.17.01 Walls and structures
04.17.02 Stacks and flue linings
04.17.03 Ancillaries, including damp courses, ties, joint reinforcement, bonding, etc.

04.18   LANDSCAPING

04.18.01 Soft landscaping
  (a) cultivating, weeding, herbicides, fertilizing, mulching
  (b) turfing, grass seeding
  (c) plants, shrubs, trees
  (d) hedges
  (e) protection

04.18.02 Hard landscaping
- (a) pavings: bricks, blocks, flags, setts, cobbles, gravel, hoggin, concrete including kerbs, edges, channels
- (b) surfaces: sports, e.g. coloured tarmacadam, sheeted, tufted, including markings and lines
- (c) walls: retaining walls, blast walls, fences, rock-filled gabions, sight and acoustic screening
- (d) fencing and hedging: fences, hedges, security fences, gates and hoardings
- (e) site/street furniture: signs, bollards, barriers, lifting barriers, seats, benches, tables, litterbins, gritbins, dustbins and planters
- (f) equipment: sports, playground, cycle stands, clothes drying
- (g) artwork: sculptures, carvings, ornamental works, flagpoles, poster hoardings, displays

04.19 ROADS

04.18.01 Sub-base
- (a) granular, soil cement, lean mix, hard core

04.18.02 Road base
- (a) macadam: wet mix, dry bound
- (b) bitumen: open texture, dense
- (c) tarmacadam: open texture, dense
- (d) tar surface

04.18.03 Surface
- (a) cold asphalt wearing course
- (b) rolled asphalt
- (c) slurry, sealing, surface dressing, bitumen spray

G+H McGill

INTERIOR CONSTRUCTION

As a multi-discipline Interior Construction Contractor we are able to offer our Clients the opportunity to place a single order for a combination of activities which make up a vital part of the overall construction programme and therefore, require close co-ordination. These activities include:

- Dry Wall Systems
- Partitions
- False Ceilings
- False Floors
- First and Second-Fix Joinery
- Passive Fire Protection
- Fire Stopping
- Floor & Wall Coverings
- Decoration
- Insulation

For further information contact:
G+H McGill Limited
McGill House
387 London Road
Hadleigh Essex
SS7 2DT
Tel: **0702 553166**
Fax: **0702 559936**
Tlx: **99280**

Member of G+H MONTAGE Group of Companies

05.00

Buildings including building services

05.01 Substructure
05.02 Superstructure
05.03 Finishings
05.04 Fittings
05.05 Services
05.06 External works

05.01 SUBSTRUCTURE

05.01.01 Site investigation
05.01.02 Site clearance/demolition
05.01.03 Removal of vegetable soil
05.01.04 Reduce level excavation
05.01.05 Basement excavation
05.01.06 Soft spots and filling
05.01.07 Sheet piling
05.01.08 Piling
05.01.09 Dewatering
05.01.10 Filling and imported filling to make up levels
05.01.11 Column bases, pile caps and ground beams
05.01.12 Strip foundations
05.01.13 Basements
05.01.14 Underpinning
05.01.15 Retaining structures
05.01.16 Rafts
05.01.17 Substructure walls
05.01.18 Floor trenches and covers
05.01.19 Underfloor drainage
05.01.20 Ground floor slab

05.02 SUPERSTRUCTURE

05.02.01 Frame and encasing
05.02.02 Upper floors
05.02.03 Roofs
05.02.04 Staircases and ramps
05.02.05 Lift shafts
05.02.06 External walls
05.02.07 Windows
05.02.08 External doors
05.02.09 Flues and chimney shafts
05.02.10 Fire escapes
05.02.11 Loading bays
05.02.12 Blast or impact-resisting requirements

05.03 FINISHINGS

05.03.01 Internal walls and finishes
05.03.02 Floor finishes
05.03.03 Ceiling finishes
05.03.04 Staircase finishes
05.03.05 Internal doors and screens
05.03.06 Internal decoration
05.03.07 External decoration
05.03.08 Special acoustic requirements
05.03.09 Fireproofing and insulation

05.04 FITTINGS

05.04.01 Fittings, cupboards, benches and counters
05.04.02 Furniture general
05.04.03 Special furniture (e.g. canteen/classroom)
05.04.04 Carpets and curtaining
05.04.05 Recreational equipment
05.04.06 Shutters, screens and blinds

05.05 SERVICES *(SEE ALSO SECTIONS 08.00, 09.00 AND 10.00)*

05.05.01 Sanitary fittings
05.05.02 Waste, soil and overflow pipes
05.05.03 Rainwater installation
05.05.04 Hot, cold and drinking water services
05.05.05 Heating installation
05.05.06 Gas installation
05.05.07 Electrical installation
05.05.08 Any engineering installation particular to the type of building
05.05.09 Telephone and telecommunications installations
05.05.10 Computer installation
05.05.11 Refuse disposal
05.05.12 Ventilation/air conditioning installation
05.05.13 Lifts and escalators
05.05.14 Overhead cranes and rails
05.05.15 Window cleaning gear
05.05.16 Special equipment (e.g. kitchen, laboratory and laundry)
05.05.17 Security and monitoring systems
05.05.18 Fire protection (e.g. alarms, sprinklers, dry risers, hose reels and extinguishers)
05.05.19 Builder's work in connection with services

05.06 EXTERNAL WORKS

05.06.01 Alterations to public highways
05.06.02 Roads
05.06.03 Parking areas/buildings
05.06.04 Paths and paved areas
05.06.05 Walling and fencing
05.06.06 Drainage
05.06.07 Lighting
05.06.08 Connection of utilities
05.06.09 Builder's work in connection with services
05.06.10 Landscaping and planting

Britains Leading Independent Structural Engineering Company

WILLIAM HARE LIMITED

Weston Street Bolton Lancashire BL3 2AT Tel 0204 26111 Telex 63277 Fax 0204 398136

06.00

Structures (see also sections 04.00 Civil engineering and 05.00 Buildings)

06.01 Supporting structures
06.02 Steelwork
06.03 Timberwork
06.04 Precast concrete work
06.05 Scaffolding

06.01 SUPPORTING STRUCTURES

06.01.01 Structures for plant and equipment including mezzanine floors, decking and any enclosing sheeting
06.01.02 Structures in support, around and access to top of plant and equipment
06.01.03 Pre-assembled units
06.01.04 Plant and equipment supports
06.01.05 Pipe racks
06.01.06 General pipe supports
06.01.07 Minor access platforms
06.01.08 Lift structures
06.01.09 Fire-fighting towers
06.01.10 Cable races
06.01.11 Unloading/loading structures (see also section 16.00)

06.02 STEELWORK

06.02.01 Supply, fabrication and erection
06.02.02 Rolled sections
06.02.03 Plates and flats
06.02.04 Hollow sections
06.02.05 Built-up box sections
06.02.06 Grillages, columns, beams, plates
06.02.07 Deck-framing portal frames
06.02.08 Trusses, trestles, towers, bracings, purlins
06.02.09 Decking, cladding, panelling
06.02.10 Walkways, platforms
06.02.11 Bridges, gantries, balconies, ramps
06.02.12 Stairways, landings, ladders, handrails, gates, cages
06.02.13 Floor plates, gratings
06.02.14 Duct covers, frames
06.02.15 On-site welding
06.02.16 Tests
06.02.17 Surface treatments including blast cleaning, flame cleaning, galvanizing, painting
06.02.18 Ancillary items including anchor bolts, holding-down bolt assemblies, packing, grouting

06.03 TIMBERWORK

06.03.01 Shoring
06.03.02 Columns, beams, trusses, portal frames
06.03.03 Trestles, towers, bracings, purlins
06.03.04 Decking, cladding, panelling, linings, sheetings, casings
06.03.05 Walkways, platforms, gantries, bridges
06.03.06 Stairways, landings, ladders
06.03.07 Handrails, balustrading

06.04 PRECAST CONCRETE WORK *(SEE SECTION 04.04.08 PRECAST CONCRETE)*

06.05 SCAFFOLDING

06.05.01 Material: steel, galvanized, alloy, other
06.05.02 Method: supply, hire, erect, maintain, adapt, dismantle, remove
06.05.03 Location: internal, external, loadbearing, below datum, in vessels
06.05.04 Type: putlog, independent, birdcage
06.05.05 Decking: boards, protection fans, safety measures
06.05.06 Ancillaries: barriers, hand and safety rails, ladders, chutes, enclosing sheeting

VAN SEUMEREN (U.K.) LTD.

Wansbeck Business Centre
Teltech House
Ashington
Northumberland
NE63 8PW

Tel. 0670 856666
Telex. 538153
telefax 0670 856582

Mobile cranes up to 1000 tons
Crawler cranes up to 600 tons
Telecopic cranes up to 400 tons
Self propelled transporters
Bundlepullers
Airbags transport
Skidding & Jacking equipment
Turnkey Project contractors

Head Office: Van Seumeren B.V. (HOLLAND)
Tel No   31-3406-95111
Fax No   31-3406-65128
Telex No 47682

Branch Office Belgium:
N. V. Seumeren
Tel No   32-91-459891
Fax No   32-91-456376
Telex 12643

Branch Office France:
Van Seumeren S.A.R.L.
Tel No   33-(1)-47885048
Fax No   33-(1)-47891869
Telex 615759

07.00

Plant including machinery and equipment

07.01 Manufacturing facilities
07.02 Process plant
07.03 Electrical generation, transmission and distribution
07.04 Process pipework

Due regard should be paid to the need for auxiliary and back up equipment.

07.01　MANUFACTURING FACILITIES

07.01.01　Manufacturing plant and equipment: because of the wide variation in types of product and facilities it is not possible to include a checklist, and the estimator must ascertain the requirements

07.02　PROCESS PLANT

07.02.01　Vessels
　　　　(a)　reactors
　　　　(b)　columns
　　　　(c)　pressure vessels
　　　　(d)　fermenters
　　　　(e)　non-pressure vessels
　　　　(f)　jacketed vessels

07.02.02 Vessel ancillary equipment
- (a) agitators and supports
- (b) internal trays
- (c) external and internal coils
- (d) packing (Raschig rings etc.)
- (e) distributors

07.02.03 Heat transfer including integral supports
- (a) fixed head exchangers
- (b) floating head exchangers
- (c) finned tube air coolers
- (d) cooling towers (see also section 04.00)

07.02.04 Process displacement equipment (section 08.00 for non-process equipment)
- (a) pumps, rotary
- (b) pumps, reciprocating
- (c) pumps, portable drum
- (d) pumps, vacuum
- (e) pumps, peristaltic
- (f) process gas compressors
- (g) fans
- (h) blowers
- (i) exhausters
- (j) ejectors and injectors

07.02.05 Compressor ancillary equipment (section 08.00 for non-process equipment)
- (a) inlet filter
- (b) inlet silencer
- (c) discharge silencer
- (d) pre-cooler
- (e) inter-cooler
- (f) after-cooler
- (g) lubricating oil tank, pump and local pipework
- (h) lubricating oil cooler
- (i) lubricating oil filter

07.02.06 Prime movers (section 08.00 for non-process equipment)
- (a) electric motors: standard, flameproof or drip proof
- (b) gas expanders
- (c) turbines: steam or gas (see also section 07.03.09)
- (d) internal combustion engines
- (e) hydraulic turbines
- (f) wind turbines

07.02.07 Filtration and separation (section 08.00 for non-process equipment)

(a) air filters
  (b) gas filters
  (c) bag filters
  (d) liquid filters
  (e) strainers
  (f) dust filters and precipitators
  (g) separators
  (h) centrifuges
  (i) cyclones
  (j) sieving and separating equipment

07.02.08 Heaters
  (a) boilers (see also section 08.00)
  (b) furnaces
  (c) ovens
  (d) kilns
  (e) non-fired heaters: induction, RF/microwave, infra-red, etc.
  (f) stacks, flues and flares (section 08.00)

07.02.09 Mixing equipment (section 08.00 for non-process equipment)
  (a) agitators
  (b) mixing and kneading machines
  (c) blending equipment

07.02.10 Disintegration equipment (section 08.00 for non-process equipment)
  (a) crushers
  (b) grinders
  (c) mills

07.02.11 Miscellaneous equipment (section 08.00 for non-process equipment)
  (a) drying equipment
  (b) chillers and scrapers
  (c) crystallizers
  (d) winterization (see also section 12.00)

07.03 ELECTRICAL GENERATION, TRANSMISSION AND DISTRIBUTION

07.03.01 Boiler
  (a) drum and internals
  (b) furnace walls
  (c) roof and enclosure

          (d)   primary superheater
          (e)   platen superheater
          (f)   final superheater
          (g)   reheater
          (h)   attemperator
          (i)   economizer
          (j)   large-bore downcomers and circulating pipework
          (k)   boiler circulating pumps including motors
          (l)   blow down, drain and vent pipework
          (m)   safety valves and silencers
          (n)   drum and unit slings
          (o)   suspension steelwork
          (p)   casings
07.03.02   Fuel firing
          (a)   mills
          (b)   feeders
          (c)   pulverized fuel/oil/gas burners
          (d)   pressurized air fans
          (e)   pulverized fuel piping
          (f)   stokers and chaingrates
          (g)   oil-firing lighting-up equipment including oil heaters
          (h)   burner management
          (i)   gas compressors
07.03.03   Air heaters
07.03.04   Sootblowers
07.03.05   Precipitators including rectifiers, transformers and local cabling
07.03.06   Draught plant
          (a)   forced draught fans and motors
          (b)   induced draught fans and motors
          (c)   flues
          (d)   ducts
          (e)   dampers
          (f)   local control and instrumentation
          (g)   insulation
          (h)   flue gas desulphurization
07.03.07   Main turbo alternator and associated items
          (a)   turbines: high pressure, intermediate pressure, low pressure
          (b)   valves: high pressure valves and governor gear, interceptor valves
          (c)   reheaters

- (d) generator including excitation system, earthing and protection
- (e) lubrication oil system and pipework
- (f) generator stator cooling system
- (g) hydrogen cooling system including storage, valves and pipework (see also section 08.00)
- (h) carbon dioxide system including storage, valves and pipework (see also section 08.00)
- (i) nitrogen system including storage, valves and pipework (see also section 08.00)
- (j) condenser/dump condenser including CW valves and tubing
- (k) feed heating system, condensate including feed regulator valves
- (l) feed pumps including main feed pump, starting and standby feed pumps, and emergency feed pumps
- (m) local control and instrumentation (if not in section 10.00)

07.03.08 Power station hydro turbines, pump turbines, alternators and motors
- (a) hydro turbine including isolating and control/governing valves with jacking and lubricating oil systems
- (b) pump turbines including isolating and control/governing valves with jacking and lubricating oil systems and declutching gear
- (c) alternators including excitation system and ancillaries
- (d) alternators/motors including excitation system and ancillaries

07.04 PROCESS PIPEWORK

07.04.01 Process pipework (section 08.00)

ledwood construction ltd

FOR THE COMPLETE SERVICE IN FABRICATION, SPECIALIST WELDING & REFINERY MAINTENANCE

CONTRACTORS TO THE OIL & PETRO-CHEMICAL ON & OFFSHORE INDUSTRIES

Waterloo Ind. Est.,

Pembroke Dock.

Pembroke (0646) 686154

Telex : 48616

Fax : (0646) 621228

MECHANICAL & CIVIL ENGINEERING

08.00

Mechanical services

08.01 Process pipework
08.02 Cooling water
08.03 Steam and condensate
08.04 Process water
08.05 Potable water
08.06 Firemains, spray and dousing
08.07 Drainage
08.08 Process effluents (solid, liquid or gaseous)
08.09 Gas and inert gas
08.10 Air
08.11 Refrigeration services
08.12 Heating, ventilating and air conditioning
08.13 Boiler plant
08.14 Chimney stacks, flues and flares
08.15 Fuel storage and preparation
08.16 Special maintenance facilities
08.17 Import/export loading facilities
08.18 Storage

This section comprises pipework and all plant and equipment ancillary to the main process.

08.01 PROCESS PIPEWORK

08.01.01 Carbon steel pipework
08.01.02 Stainless steel pipework
08.01.03 Other pipework
08.01.04 Undesigned small-bore pipework
08.01.05 Pipe fittings
08.01.06 Valves: motorized valves, control valves and relief valves
08.01.07 In-line equipment
08.01.08 Internal lining
08.01.09 Insulation (see also section 12.00)
08.01.10 External coating (see also section 13.00)

08.01.11 Non-destructive testing
08.01.12 Testing and cleaning
08.01.13 Trace heating (see also section 09.00)

08.02 COOLING WATER

08.02.01 Water source and supply (see also section 04.00) including pumps and motors, and screens including backwash
08.02.02 Pre-treatment storage tanks or ponds (see also section 04.00)
08.02.03 Treatment
- (a) screens and filters
- (b) flocculators
- (c) settling tanks
- (d) desalinization
- (e) demineralizers
- (f) antibiological, bactericidal and fungicidal
- (g) antiscaling
- (h) de-aerators
- (i) reverse osmosis plants
- (j) storage, handling and dosing of treatment chemicals

08.02.04 Post-treatment storage tanks or ponds (see also section 04.00)
08.02.05 Distribution
- (a) circulating and purge pumps and motors including gearboxes
- (b) headers, pipework and manifolds (flow and return)
- (c) trunk mains (flow and return)
- (d) undesigned small-bore pipework
- (e) isolating valves including actuators
- (f) internal lining
- (g) insulation and external coating (see also sections 12.00 and 13.00)
- (h) non-destructive testing
- (i) testing and cleaning

08.02.06 Treatment for recycling or return
- (a) cooling towers
- (b) spray ponds
- (c) filters
- (d) chlorination plant

08.03 STEAM AND CONDENSATE *(FOR RELEVANT ITEMS SEE SECTIONS 08.02.01 FOR WATER SOURCE, AND 08.02.02–08.02.04 FOR TREATMENT AND STORAGE)*

08.03.01 Boiler plant (see sections 08.13 for boiler plant, 08.15 for fuel storage, and 08.14 for stacks and flues)
08.03.02 Steam accumulator
08.03.03 Distribution pipework
    (a) pipework trunk mains
    (b) trace-heating pipework and undesigned small-bore pipework
    (c) isolating, reducing and relief valves
    (d) traps, strainers, separators and manifolds
    (e) expansion devices
    (f) internal lining
    (g) insulation and external coating (see also sections 12.00 and 13.00)
    (h) non-destructive testing
    (i) testing and cleaning
08.03.04 Condensate
    (a) collecting and flash tanks
    (b) pumps and motors
    (c) pipework
    (d) undesigned small-bore pipework
    (e) headers and manifolds
    (f) isolating valves
    (g) condensate polishing
    (h) insulation and external coating (see also sections 12.00 and 13.00)
    (i) non-destructive testing
    (j) testing and cleaning
    (k) for feed heating see section 07.03.08 on hydro turbines
08.03.05 High-pressure work: high-pressure steam feeds such as that between boiler and turbine in power generation

08.04   PROCESS WATER *(FOR RELEVANT ITEMS SEE SECTIONS 08.02.01 FOR WATER SOURCE, AND 08.02.02– 08.02.04 FOR TREATMENT AND STORAGE)*

08.04.01   Boiler plant (see sections 08.13 for boiler plant, 08.15 for fuel storage, and 08.14 for stacks and flues)
08.04.02   Distribution pipework
  (a)   pump and motors
  (b)   pipework mains and distribution spurs
  (c)   undesigned small-bore pipework
  (d)   valves, motorized valves, control and relief valves
  (e)   in-line equipment
  (f)   internal lining
  (g)   insulation and external coating (see also sections 12.00 and 13.00)
  (h)   non-destructive testing
  (i)   testing and cleaning

08.05   POTABLE WATER *(SECTION 08.02 FOR LIST OF ITEMS AS APPLICABLE)*

08.06   FIREMAINS, SPRAY AND DOUSING
CHECK THAT BELOW-GROUND FIREMAIN PIPEWORK IS INCLUDED IN SECTION 04.00 CIVIL ENGINEERING, TOGETHER WITH STORAGE PONDS AND/ OR INTAKES; FOR ABOVE-GROUND PIPEWORK SEE SECTION 08.02 FOR LIST OF ITEMS AS APPLICABLE

08.06.01   Pressure storage tanks
08.06.02   Compressors including motors
08.06.03   Pumps and motors
08.06.04   Diesel pumps
08.06.05   Foam systems
08.06.06   Dry risers
08.06.07   Sprinklers and monitors
08.06.08   Detectors
08.06.09   Carbon dioxide systems
08.06.10   Halon systems

08.07 DRAINAGE: *CHECK THAT BELOW-GROUND SURFACE WATER, OILY WATER AND FOUL SEWER DRAINAGE ARE INCLUDED IN SECTION 04.00, CIVIL ENGINEERING, TOGETHER WITH SEPARATORS, SETTLEMENT TANKS, SEWAGE TREATMENT FACILITIES AND OUTFALLS; FOR ABOVE-GROUND PIPEWORK SEE SECTION 08.02 FOR LIST OF ITEMS AS APPLICABLE*

08.08 PROCESS EFFLUENTS (SOLID, LIQUID OR GASEOUS)

08.08.01 Collection and storage
  (a) solids collection system
  (b) effluent drainage systems, sumps, ponds, interceptors and tanks
  (c) fume intake ducts and fans
  (d) special handling of contaminated/toxic waste
  (e) ash handling

08.08.02 Treatment
  (a) neutralizers
  (b) bacteriological beds
  (c) activated sludge beds
  (d) oxidation/reduction
  (e) heavy metals removal
  (f) chemical treatment
  (g) deodorizers
  (h) absorbers
  (i) chemicals storage and handling including air supply for aerators

08.08.03 Separation
  (a) flocculators
  (b) settling tanks and beds
  (c) drainage beds
  (d) concentrators
  (e) cyclones and hydrocyclones
  (f) filters
  (g) grit arrestors
  (h) electrostatic precipitators

08.08.04 Waste storage and handling

08.08.05 Disposal
- (a) incinerators
- (b) chimney and stacks
- (c) dumping costs
- (d) contaminated/toxic waste
- (e) outfalls: sewer connections and tidal outfall control
- (f) packaging/selling

08.08.06 Distribution
- (a) pumps and motors
- (b) interconnecting pipework
- (c) undesigned small-bore pipework
- (d) internal lining
- (e) insulation and external coating (see also sections 12.00 and 13.00)
- (f) non-destructive testing
- (g) testing and cleaning

08.09 GAS AND INERT GAS *(EXCLUDING AIR)*

08.09.01 Source

08.09.02 Unloading and storage
- (a) handling: HP containers
- (b) receivers

08.09.03 Generators
- (a) HP containers
- (b) generators
- (c) evaporators

08.09.04 Treatment
- (a) filters
- (b) driers
- (c) heaters
- (d) refrigerators

08.09.05 After-treatment and storage
- (a) HP, LP

08.09.06 Distribution
- (a) compressors including motors, coolers and ancillaries
- (b) headers and manifolds
- (c) pipework and isolating valves
- (d) undesigned small-bore pipework
- (e) reducing sets

(f) insulation and external coating (see also sections 12.00
(g) non-destructive testing
(h) testing and cleaning

8.10 AIR *(INCLUDING COMPRESSED AIR)*

08.10.01 Air intake
08.10.02 Pre-treatment
    (a) filters
    (b) washers
    (c) silencers
08.10.03 Compressor plant
    (a) compressors including motors, coolers and ancillaries
08.10.04 After treatment
    (a) filters including sterilizing filters
    (b) scrubbers including sterilizing scrubbers
    (c) electrostatic precipitators
    (d) humidifiers and dehumidifiers
    (e) heaters/coolers
08.10.05 Storage
    (a) air receivers
    (b) stand-by cylinders
08.10.06 Distribution
    (a) headers and manifolds
    (b) pipework and isolating valves
    (c) undesigned small-bore pipework
    (d) traps, separators, etc.
    (e) reducing valves
    (f) special outlets (e.g. for breathing hoods, air-driven tools, etc.)
    (g) insulation and external coating (see also sections 12.00 and 13.00)
    (h) non-destructive testing
    (i) testing and cleaning

08.11 REFRIGERATION SERVICES

08.11.01 Storage
    (a) refrigerant including gas cylinder handling
    (b) cooling-medium raw materials

08.11.02 Cooling medium make-up and storage
- (a) make-up tanks
- (b) solids handling
- (c) reservoir tanks

08.11.03 Refrigeration plant
- (a) compressors: mechanical or jets
- (b) absorbers
- (c) freezing mixture tanks
- (d) coolers/condensers
- (e) receivers
- (f) evaporators and coils
- (g) auxiliaries including eliminators (moisture and oil)
- (h) control gear

08.11.04 Refrigerant distribution
- (a) pumps and motors
- (b) HP headers and manifolds
- (c) HP trunk mains and isolating valves
- (d) LP headers and manifolds
- (e) LP trunk mains and isolating valves
- (f) undesigned small-bore pipework

08.11.05 Cooling medium distribution
- (a) pumps and motors
- (b) headers and manifolds
- (c) trunk mains and isolating valves
- (d) undesigned small-bore pipework
- (e) insulation and external coating (see also sections 12.00 and 13.00)
- (f) non-destructive testing
- (g) testing and cleaning

08.12 HEATING, VENTILATING AND AIR CONDITIONING

08.12.01 Main plant
- (a) air fans including motors
- (b) pumps including motors
- (c) refrigeration units
- (d) heating units
- (e) heat exchangers including calorifiers
- (f) biological controls
- (g) humidifiers

        (h)   filters
        (i)   cooling towers
        (j)   local pipework
        (k)   drains
        (l)   storage tanks
        (m)   local ductwork, silencers, intakes, exhausts
        (n)   dampers including actuators
        (o)   local controls (see also section 10.00)
        (p)   local cabling (see also section 09.00)
        (q)   insulation and external coating (see also sections 12.00 and 13.00)
        (r)   control panels (see also section 10.00)

08.12.02 Ductwork
        (a)   flow duct system
        (b)   return duct system
        (c)   registers
        (d)   spray/steam injectors
        (e)   dampers including actuators
        (f)   insulation and external coating (see also sections 12.00 and 13.00)

08.12.03 Hot water systems
        (a)   flow and return pipework
        (b)   manifolds
        (c)   radiators including fan assisted
        (d)   local circulating pumps including motors
        (e)   valves
        (f)   insulation and external coating (see also sections 12.00 and 13.00)
        (g)   testing and cleaning

08.12.04 Direct acting and local system
        (a)   local air-conditioning units
        (b)   electric heaters
        (c)   gas/liquified petroleum gas (LPG)/oil heaters

08.12.05 Sensors and controls (see also section 10.00)
        (a)   temperature sensors
        (b)   local area controls
        (c)   humidity sensors
        (d)   radiation sensors

08.12.06 Control system (see also section 10.00)
        (a)   remote control panels
        (b)   system alarms

08.12.07 Cabling (see also section 09.00)
        (a)   power cables

          (b)   main plant cables
          (c)   damper/valve actuators cabling
          (d)   control cables
          (e)   sensor control cables
          (f)   actuator/valve control cables
          (g)   remote stations to main panel

08.12.08   Pipework
          (a)   hot water trunk mains and distribution
          (b)   gas/LPG/oil
          (c)   refrigerant
          (d)   system set-up, testing and commissioning

08.13   BOILER PLANT

08.13.01   Boilers including waste heat and electrode boilers
08.13.02   Burners
08.13.03   Stokers
08.13.04   FD and ID fans and motors
08.13.05   Superheaters
08.13.06   De-superheaters
08.13.07   Economizers
08.13.08   Air heaters and motors
08.13.09   Acid or chemical cleaning
08.13.10   Sootblowers
08.13.11   Shot cleaners
08.13.12   Grit arrestors
08.13.13   Precipitators, bag filters, etc.
08.13.14   Ash handling
08.13.15   Control gear
08.13.16   Blowdown
08.13.17   Ductwork and dampers including actuators

08.14   CHIMNEY STACKS, FLUES AND FLARES

08.14.01   Lining, flues and windshields
08.14.02   Cladding
08.14.03   Anti-corrosion treatment
08.14.04   Lightning conductors
08.14.05   Warning lights
08.14.06   Flare igniters and knock-out drums

08.15 FUEL STORAGE AND PREPARATION

08.15.01 Storage areas, bunkers, tanks, gas holders
08.15.02 Heating and pumping plant
08.15.03 Pulverizing plant if not with boiler
08.15.04 Solid fuel handling and weighing

08.16 SPECIAL MAINTENANCE FACILITIES:
FOR ANY ITEM WITHIN SECTION 08.00 AS REQUIRED

08.16.01 Winterization (see also section 12.00)
08.16.02 Cranes, hoists, runway beams, etc.

08.17 IMPORT/EXPORT LOADING FACILITIES:
ALLOW FOR ANY ITEMS HERE ONLY IF NOT INCORPORATED WITHIN SECTION 16.00

08.17.01 Weighing devices, weighbridges, load cells, etc.
08.17.02 Cranes
08.17.03 Fork-lift trucks and specialist vehicles/attachments
08.17.04 Special rail truck handling facilities
08.17.05 Unloading and loading arms including screw unloaders
08.17.06 Barge tipplers
08.17.07 Conveyors, chutes and silos including magnetic separators, gauging, etc.
08.17.08 Vacuum or hydraulic transportation plants
08.17.09 Vacuum cleaning systems
08.17.10 Can, drum, bottling, packet filling, emptying and handling devices
08.17.11 Boom stackers, bucket wheel store/reclaim machines
08.17.12 Batching plants

08.18 STORAGE

08.18.01 Storage tanks, vessels or spheres (see also section 14.05)
08.18.02 Skimming equipment to settling tanks

AOC International Ltd
Alba Gate
Stoneywood Park
Dyce
Aberdeen AB2 0HN
(0224) 770033

Mechanical & Electrical construction, installation and maintenance

----

Onshore and Offshore

Boulting

Electrical & Instrumentation Contracting
and Switchgear Manufacturers

COMMERCIAL AND INDUSTRIAL ELECTRICAL
AND INSTRUMENTATION INSTALLATIONS
PROJECTS £10,000 to £5 MILLION
UNDERTAKEN

PROJECT MANAGEMENT & CONSTRUCTION
FOR BUILDING SERVICES

- PROGRAMMED MAINTENANCE
- DESIGN SERVICE
- M & E PACKAGES

- CONTACT DAVID TRANTER
MEMBERS OF THE ECA, NICEIC, JIB

Tel: (0925) 726661

THE POWER TO SERVE

SERVING WARRINGTON SINCE 1918

A MEMBER OF THE
CONTRACTING DIVISION,
THOMAS ROBINSON
GROUP PLC

09.00

Electrical services

09.01 EHV system (incoming supplies above 11 kV)
09.02 HV system (exceeding 440 V up to 11 kV)
09.03 MV system (440 V and below)
09.04 DC system
09.05 Transformers
09.06 Lighting, heating and small power
09.07 Emergency generation
09.08 Miscellaneous equipment
09.09 Cabling
09.10 Testing and commissioning
09.11 Civil aspects

Cabling: where cabling is referred to, this includes as appropriate, a supply of cable together with drum allowances, tails and draw wires/cables, pulling/laying, glanding, terminating, cleating, bonding, identification labels, lug/ferrule fitting and connecting sleeves. It also includes the supply and fitting of cabling ancillaries, glands, crimps, ferrules, clamps, ties, labels and through joints. Account needs to be taken of control cables (multi-core or multi-pair) and trunk cabling between junction boxes/kiosks, control stations, relay/equipment room, etc. Local and/or dedicated cabling may be taken with the item or with the general cabling. Control stations, alarm fascias, emergency stops, bell pushes, if not included in Section 10.00 Instrumentation and controls, should also be taken into account.

Hazardous areas (seismic, flameproof, dustproof etc.): allowance should be made for seismic requirements, flameproofing, dustproofing, and/or special hazardous conditions, as appropriate for both equipment and installation.

09.01 EXTRA HIGH VOLTAGE (EHV) SYSTEM (INCOMING SUPPLIES ABOVE 11 kV)

09.01.01 EHV switchgear and isolators including foundations and compounds (if not separate), protective gear, metering, local

09.01.02	EHV cable (if not general cabling) including terminations and structure, sealing ends, watersheds, co-ordinating gaps, through joints, oil/gas associated systems and connections, droppers, flexible connections and jumpers
09.01.03	EHV overhead lines including towers/poles and their foundations, cross arms, guys, insulators, surge diverters, line fuses, auto-reclosers, conductors, spacers and dampers, earth wires, droppers, flexible connections and jumpers
09.01.04	EHV transformers (EHV/HV) including foundations and compounds (if not separate), bushings and watersheds, tap changers, coolers, fans, motors, control gear, interconnecting pipework, local cubicles, oil fill, together with relay panels, protective gear and associated current transformers
09.01.05	Instrument transformers (where not supplied elsewhere) (a) voltage including, if appropriate, line couplers, line traps, support structures, local cubicles and local wiring (b) current including support structures, local cubicles and local wiring
09.01.06	Supply authority capital contribution
09.01.07	Supply authority maintenance charges
09.01.08	Charges for training and authorization for switching

09.02 HIGH VOLTAGE (HV) SYSTEM (EXCEEDING 440 V UP TO 11kV)

09.02.01	HV switchgear including foundations/support and compounds (if not separate), protective gear, metering, local controls, earthing devices, key cabinets (including contents), interlocking and exchange boxes
09.02.02	HV cables (if not general cabling) including terminations, sealing ends, watersheds, co-ordinating gaps, through joints, lugs, glands, sealing and earth links
09.02.03	HV overhead lines including towers/poles and their foundations, cross arms, guys, insulators, surge diverters, line fuses, auto-reclosers, conductors, dampers, earth wires, droppers, flexible connections and jumpers
09.02.04	HV transformers (HV/HV) (HV/MV) including foundations and compounds (if not separate), bushings and watersheds, tap-changers, coolers, fans, motors, control gear, interconnecting

	pipework, local cubicles, oil fill, together with relay panels, protective gear and associated current transformers (NB: HV/MV integral transformers are included with MV switchboards – see section 09.03.01)
09.02.05	Earthing transformers
09.02.06	Earthing resistors including fill, heaters and thermostats

09.03 MEDIUM VOLTAGE (MV) SYSTEM (440 V AND BELOW)

09.03.01 MV switchboards, control gear, motor starters including foundations/support and compounds (if not separate), integral transformers, busbars, outgoing circuit breakers, contactors, motor control gear, mcbs, isolators, protective gear, metering, local switchboard-mounted selector switches, control buttons, auxiliary relays, gland plates, earthing bars and links and key cabinets (including contents)

09.03.02 MV fuseboards, distribution boards and feeder pillars including case, fuseways with fuses and gland plates

09.03.03 MV outgoing circuits
  (a) switchboard, contactor, motor control or fuseway as appropriate
  (b) power cable
  (c) control cables (600/1000 V grade) to local control station and/or emergency stop
  (d) control cables (100 V grade) to local junction box (excluding junction box) to sensors, transducers, (excluding devices, detectors, etc.) (see also section 10.00 Instrumentation and controls)
  (e) earthing of individual circuit items if appropriate.

09.04 DC SYSTEM

09.04.01 Storage batteries including supports, acid fill, inter-cell connections, and accessories
  (a) 240 V
  (b) 110 V
  (c) 50 V
  (d) Other

09.04.02   Chargers including foundations/support (if not separate)
09.04.03   DC switchgear including foundations/support and compounds (if not separate) protective gear, metering, local controls, earthing devices, key cabinets (including contents) interlocking and exchange boxes and local cables
09.04.04   DC fuseboards including case, fuseways with fuses and gland plates
09.04.05   Rectifiers including foundations/support/mountings, interconnections, terminals, cooling, fuse/protection and metering
09.04.06   Inverters and uninterruptible power supplies including foundation/support/mountings, protective gear, metering, control gear, earthing devices, key cabinets (including contents) and interlocking
09.04.07   DC cables
09.04.08   Outgoing DC circuits
   (a)   circuit breaker, contactor, fuseway or motor control device as appropriate
   (b)   power cable
   (c)   control cables (600/1000 V grade) to local control station and/or emergency stop
   (d)   control cables (100 V grade) to local junction box (excluding junction box) to sensors, transducers, (excluding devices, detectors, etc.) (see also section 10.00 Instrumentation and controls)
   (e)   earthing of individual circuit items if appropriate, emergency lighting and DC power and plant items

09.05   TRANSFORMERS: *TRANSFORMERS NOT INCLUDED ELSEWHERE AND/OR OF A SPECIALIZED NATURE INCLUDING FOUNDATIONS AND COMPOUNDS (IF NOT SEPARATE), BUSHINGS AND WATERSHEDS, TAP CHANGERS, COOLERS, FANS, MOTORS, CONTROL GEAR, INTERCONNECTING PIPEWORK, LOCAL CUBICLES, OIL FILL, TOGETHER WITH RELAY PANELS, PROTECTIVE GEAR AND ASSOCIATED CURRENT TRANSFORMERS*

09.05.01   Generator transformers
09.05.02   Generator unit transformers
09.05.03   Generator earthing transformers including, in addition to general note above, kick fuses, grid resistors, connections to generator terminals including insulators

09.05.04 Station/auxiliary transformers
09.05.05 Instrument transformers where separately supplied for metering and/or protection including mountings
    (a) voltage transformers
    (b) current transformers
    (c) auxiliary current transformers
09.05.06 Portable transformers
09.05.07 Transformer oil-conditioning equipment

09.06 LIGHTING, HEATING AND SMALL POWER

09.06.01 Lighting equipment including fittings, lighting panels, luminaires, switches, starters, lamps and cabling (if not separate)
    (a) plant
    (b) buildings
    (c) street and area lighting including towers/poles, standards, columns
    (d) flood lighting including towers/poles
    (e) high bay lighting
    (f) emergency lighting
    (g) exit lighting
    (h) traffic lights and road signs
    (i) navigation lights
    (j) airfield lighting including control equipment, regulators, isolating transformers, plugs and sockets
    (k) any other listed item
09.06.02 Switchboards, fuseboards, control panels
09.06.03 Power sockets
09.06.04 Welding sockets
09.06.05 Heaters, boilers including thermostats, control switches, relays and contactors
    (a) heaters
    (b) water heaters
    (c) electrode boilers
09.06.06 Trace heating (if not part of plant controls) including heating tapes, thermostats, local connecting boxes, cabling, insulation and finishing (if not separate)
09.06.07 Clocks including master if appropriate, slaves, outlet boxes and dedicated cabling
09.06.08 Fans including control devices

09.06.09 Security systems including power units, detectors, sensors, closed circuit TV cameras, control and display units, dedicated floodlights, bells, sirens and dedicated cabling
09.06.10 Card access system including readers, control units/computers, door/gate/turnstile control gear (including door/gate/turnstile if not included elsewhere), dedicated cabling and card issuing devices
09.06.11 Staff recording system including clocks, card readers, storage racks, control units/computers and display/output devices
09.06.12 Building monitoring system including energy, lighting, fire alarm and security systems in one package, and dedicated cabling
09.06.13 Any other listed item

09.07 EMERGENCY GENERATION *(FOR MAIN GENERATING EQUIPMENT SEE SECTION 07.03)*

09.07.01 Prime mover (diesel, petrol, gas, gas-turbine, hydro) including foundations, controls, local pipework
09.07.02 Alternator including excitation system, barring gear if appropriate
09.07.03 Fuel storage system
09.07.04 Lubrication system
09.07.05 Cooling systems
 (a) prime mover
 (b) oil coolers
 (c) storage
 (d) treatment
09.07.06 Neutral and earthing system
09.07.07 Synchronizing, protection, automatic control

09.08 MISCELLANEOUS EQUIPMENT

09.08.01 Power factor correction capacitors
09.08.02 Load-shedding equipment

09.09 CABLING *WHERE NOT ALREADY INCLUDED AS PART OF OTHER WORKS*

09.09.01 Generator main connections complete, including own ventilation/cooling system, local controls and cubicles and type testing if appropriate

09.09.02 Power cables
    (a) EHV
    (b) HV: 11 kV, 6.6 kV and 3.3 kV
    (c) MV 415/240 including the systems of section 09.03 and 09.06 as appropriate
    (d) DC system including the systems of section 09.04 as appropriate

09.09.03 Control and communication system cables
    (a) 600/1000 V grade (110 V system)
    (b) 100 V grade (light current systems)
    (c) screened cable
    (d) radio
    (e) special cables, thermocouple tails, etc.
    (f) other cables

09.09.04 MICC cable including olives, pots, sealing compound, beads/sleeves, saddles

09.09.05 Supports
    (a) racks
    (b) trays
    (c) conduit
    (d) trunking
    (e) transits for fireproofing
    (f) support steelwork for above, and control stations

09.09.06 Ancillary items including gland plates, internal terminals, links jumpering, ferrules, sleeves
    (a) junction boxes
    (b) marshalling kiosks

09.09.07 Earthing system
    (a) earth mat/plates
    (b) earth rods, including driving
    (c) earth tape
    (d) trenches and cable connections
    (e) connections of plant structure to earthing system and test clamps
    (f) earthing of pylons, towers, poles, posts and fences
    (g) radio frequency earthing and system

09.09.08 Fireproofing of cable
09.09.09 Cathodic protection including anodes, pits, power supply, cabling and fittings
09.09.10 Lightning protection including conductors, tape/cable, cleats, earth electrodes and links
09.09.11 Pylons, towers, poles, posts including foundations, catenaries, and all fittings

---

09.10 TESTING AND COMMISSIONING: *SITE TESTING, PRE-COMMISSIONING AND COMMISSIONING COMPONENTS AND SYSTEMS AS REQUIRED WHERE NOT ALLOWED FOR ELSEWHERE*

09.11 CIVIL ASPECTS: *IF CIVIL WORKS DO NOT COVER, ACCOUNT IS TO BE TAKEN OF PRECAST TRENCHES, EXCAVATION, SAND, CABLE TILES, ROUTE MARKERS, BACKFILL, COVERS, CONDUITS, PREFORMED DUCTS, SEALING, WEATHER PROTECTION AND GENERAL BUILDER'S WORK*

10.00

Instrumentation and controls including communications

10.01 Computer based systems
10.02 Electronic/relay based systems
10.03 Pneumatic systems
10.04 Hydraulic systems
10.05 Process input devices, transducers, transmitters
10.06 Output devices, VDUs, indicating instruments
10.07 Control desks and panels
10.08 Electronic/electrical ancillaries
10.09 Power packs
10.10 Fire protection
10.11 Remote viewing and closed circuit TV
10.12 Electrical actuators for valves, dampers and controllers
10.13 Instrument valves, blowdown racks, impulse piping
10.14 Communications systems
10.15 Data communications
10.16 Portable instruments
10.17 Applications software
10.18 Manuals
10.19 Cabling
10.20 Instrument piping
10.21 Testing and commissioning
10.22 Control valves
10.23 Civil aspects

For cabling and hazardous areas see also footnote to section 09.00.

10.01 COMPUTER BASED SYSTEMS

10.01.01 Computers including consoles/cabinets/housings, supports/ mounts, power supplies, integral ventilating/cooling, dedicated inter-unit cabling, highway connections if appropriate,

input/putput devices, displays, data storage media, operating systems
- (a) main computer
- (b) back-up computer
- (c) distributed computers
- (d) programmable logic controllers
- (e) data loggers and acquisitions
- (f) simulators

10.01.02 Peripherals (items not included within a computer system as in section 10.02.01)
- (a) desk/keyboard/display unit/software loading
- (b) keyboards
- (c) VDU
- (d) printers
- (e) plotters
- (f) recorders
- (g) back-up devices/tape streamer

10.01.03 Ancillaries
- (a) dedicated cabling

10.01.04 Support services (where not included in other sections of the estimate)
- (a) heating and ventilating/cooling system
- (b) power supplies

10.02 ELECTRONIC/RELAY-BASED SYSTEMS

10.02.01 Equipment cubicles including cubicle, supports/mounts, cable gland plates, termination, instrument pipework, relays, interfaces, amplifiers, logic units, transmitters, receivers, all internal wiring and piping, self-contained power supplies, fuses, indicating lights, self-contained ventilation cooling systems, anti-condensation heaters
- (a) sequence control
- (b) interlock
- (c) valve selection
- (d) alarm/annunciator
- (e) modulating controls

10.03 PNEUMATIC SYSTEMS *INCLUDING CABINETS/HOUSINGS, SUPPORTS/MOUNTINGS, INTEGRAL AIR COMPRESSORS, LOCAL PIPING*

10.03.01 Actuation systems
    (a) electrical/pneumatic converters
    (b) actuators
10.03.02 Self-contained systems: in addition to generic inclusion, detectors, transducers, setting devices, actuator/drive and controlling devices (valves etc.)

10.04 HYDRAULIC SYSTEMS

10.04.01 Actuation system including cabinets/housings, supports/ mounts, integral pumps, reservoirs, tanks, piping and any converters

10.05 PROCESS INPUT DEVICES, TRANSDUCERS, TRANSMITTERS

10.05.01 Transducers (and integral transmitters) including housings/ mountings/supports, weatherproofing, environmental protection, internal/local wiring, internal power packs, if appropriate
    (a) electrical
    (b) pressure
    (c) temperature (including cold junction box if applicable)
    (d) level
    (e) flow
    (f) ionizing radiation, X-ray and nuclear
    (g) position, speed, acceleration and seismic (including anemometer)
    (h) mass
    (i) humidity
    (j) chemical composition
    (k) optical
    (l) other listed items
10.05.02 Transmitters (if separate)
10.05.03 Local control stations

(a) push button stations and selectors
(b) emergency stops
(c) multi-function local controllers
(d) support steelwork for above

10.06 OUTPUT DEVICES, VDU'S, INDICATING INSTRUMENTS

10.06.01 Visual display units (when not part of complete system)
10.06.02 Mimic diagrams
10.06.03 Indicating instruments
10.06.04 Electrical integrating instruments
10.06.05 Mechanical integrating instruments
10.06.06 Recorders
10.06.07 Alarm fascias, annunciators

10.07 CONTROL DESKS AND PANELS *WHERE NOT INCLUDED AS PART OF A COMPREHENSIVE CONTROL SCHEME, BUT INCLUDING AS APPROPRIATE: MOUNTINGS/SUPPORTS, ENVIRONMENTAL PROTECTION, WEATHERPROOFING, ALL INTERNAL WIRING AND PIPING, VENTILATION, HEATING, DRAINS, TERMINALS/CONNECTIONS/SOCKETS FOR EXTERNAL CABLING, PIPE CONNECTORS FOR EXTERNAL PIPING*

10.07.01 Desks
    (a) process control
    (b) electrical system
    (c) cooling water control
    (d) ancillary plant list
10.07.02 Control panels
10.07.03 Instrument panels

10.08 ELECTRONIC/ELECTRICAL ANCILLARIES *INCLUDING HOUSING, MOUNTINGS/ SUPPORTS, INTERNAL POWER PACKS, INTERNAL WIRING, ENVIRONMENTAL PROTECTION, WEATHERPROOFING, VENTILATION, COOLING, HEATING, DRAINS AND TERMINALS/CONNECTIONS/ SOCKETS FOR EXTERNAL CABLING*

10.08.01 Signal conditioning equipment
10.08.02 Three-term controllers
10.08.03 Additional items, e.g. special additional limit switches

10.09 POWER PACKS

10.09.01 Electrical power supply units: separate power supply units for providing AC and/or DC supplies to equipment without integral power units
    (a) instrument supplies
    (b) computer supplies
    (c) auxiliary supplies
    (d) other listed items
10.09.02 Instrument and control air compressor including baseplates, housings, support/mountings, environmental protection, weatherproofing, all self-contained piping and wiring, drains, driers, receivers, integral instruments, terminal blocks and pipe connectors for external connections
10.09.03 Hydraulic power packs including baseplates, housings, support/mountings, environmental protection, weatherproofing, storage tanks, all self-contained piping and wiring, drains, integral instruments, terminal blocks and pipe connectors for external connections

10.10 FIRE PROTECTION *INCLUDING SENSORS, CENTRAL AND ZONE MONITORS, DEDICATED CABLING, INTEGRAL POWER SUPPLIES, BUT EXCLUDING WATER/ FOAM/GAS PIPEWORK AND NOZZLES, AND EXCLUDING DIRECT-ACTING SPRINKLER SYSTEM DETECTOR INCLUDED WITHIN 08.06*

10.10.01 Sensors
10.10.02 Zone monitors and relays
10.10.03 Central monitor

10.11 REMOTE VIEWING AND CLOSED CIRCUIT TV

10.11.01 Cameras including all mountings, housings, supports, environmental protection and weatherproofing, and all facilities for swivelling, tilting and zooming
10.11.02 Monitors
10.11.03 Dedicated cabling
10.11.04 Ancillary equipment
10.11.05 Control equipment
10.11.06 Video recorders
10.11.07 Lighting gear and supplies

10.12 ELECTRICAL ACTUATORS FOR VALVES, DAMPERS AND CONTROLLERS

10.12.01 Valve actuators
    (a) isolating/regulating actuators
    (b) modulating actuators
10.12.02 Damper actuators
10.12.03 Slave actuators for controllers

10.13 INSTRUMENT VALVES, BLOWDOWN RACKS, IMPULSE PIPING *INCLUDING ISOLATING AND TEST VALVES AND AIR RELEASE FACILITIES*

10.13.01 Instrument valves
10.13.02 Instrument blowdown racks including drain facilities to appropriate terminal point
10.13.03 Instrument impulse piping
10.13.04 Manifolds

10.14 COMMUNICATIONS SYSTEMS

10.14.01 PABX/PBX system including exhange equipment, main distribution frame, sockets, operator's consoles, test equipment and dedicated cabling
10.14.02 Telephone instruments including acoustic enclosures if appropriate
10.14.03 Public address systems including cabinets, audio modules, amplifiers, microphones, control unit, speakers and dedicated cabling
10.14.04 Staff location/paging systems including transmitter, aerial, encoders, transceivers, battery-charging rack and dedicated cabling
10.14.05 Direct wire system including exchange equipment, instruments and dedicated cabling
10.14.06 Telex including machines, dedicated cabling, exchange equipment
10.14.07 Facsimile including machines, dedicated wiring and ancillaries
10.14.08 Modems and computer interfaces including equipment, dedicated wiring, power supplies and controlling software
10.14.09 Alarm/bell system including bell pushes, bells, sirens, repeaters, auto-callers, dedicated wiring and power supplies
10.14.10 Radio system including transmitter, receiver, field and/or portable receivers/transceivers, aerials, connecting radio cables, control station, battery charging racks
10.14.11 Intercom system including master station equipment, outstations and dedicated wiring

10.15 DATA COMMUNICATIONS

10.15.01 Modems
10.15.02 Multiplexers
10.15.03 Data links
10.15.04 Dedicated cabling
10.15.05 Dedicated power supplies
10.15.06 Other interfaces
10.15.07 Control devices
10.15.08 Control software
10.15.09 External charges

10.16 PORTABLE INSTRUMENTS *INCLUDING ALL CASES, ACCESSORIES, LEADS, PROBES, CONNECTORS, TAPPING DEVICES, HOSES, DRAINS, VESSELS, ETC.*

10.16.01 Electrical/electronic instruments
 (a) volt/ammeters
 (b) wattmeters
 (c) electrostatic instruments
 (d) oscilloscopes
 (e) oscillator/signal generators
 (f) frequency meters
 (g) pyrometer/thermocouple
 (h) other listed items
10.16.02 Pneumatic/hydraulic instruments
 (a) pressure gauges
 (b) level gauges
 (c) flow meters
 (d) other listed items

10.17 APPLICATIONS SOFTWARE

10.17.01 Software
10.17.02 Computer operator manuals

10.18 MANUALS

10.18.01　Operating manuals
10.18.02　Maintenance manuals
10.18.03　Spares manuals
10.18.04　Testing and commissioning manuals

10.19 CABLING *WHERE NOT ALREADY INCLUDED AS PART OF OTHER WORKS OR WITHIN SECTION 09.09*

10.19.01　Power cables
  (a)　MV 415/240 V including systems of section 09.03 and 09.06 as appropriate
  (b)　DC system including system of section 09.04 as appropriate
10.19.02　Control and communication system cables
  (a)　600/1000 V grade (110 V system)
  (b)　100 V grade (light current systems)
  (c)　screened cables
  (d)　radios
  (e)　special cables, thermocouple tails, etc.
  (f)　other cables
10.19.03　MICC cables including olives, pots, sealing compound, beads/sleeves, saddles
10.19.04　Supports for pipework where appropriate as well as cables
  (a)　racks
  (b)　trays
  (c)　conduits
  (d)　support steelwork for above, and control stations
10.19.05　Ancillary items including internal terminals, links, jumpering, ferrules, sleeves
  (a)　junction boxes
  (b)　marshalling kiosks
  (c)　Transits for fireproofing
10.19.06　Earthing system
  (a)　earth mat/plates
  (b)　earth rods including driving
  (c)　earth tape
  (d)　connections of plant structure to earthing system

(e) earthing of pylons, towers, poles, posts and fences
(f) radio frequency earthing and system
(g) trenches and cable connections

10.19.07 Fireproofing of cable
10.19.08 Pylons, towers, poles, posts including foundations, all fittings

10.20 INSTRUMENT PIPING, TUBING, MULTI-CORE TUBING AND ASSOCIATED VALVES, MANIFOLDS AND FITTINGS *INCLUDING SUPPORTS, RACKS, TRAYS, NECESSARY TERMINAL UNIONS, JOINTS, INTERMEDIATE UNIONS, CLEATS, CLIPS, INSULATION, WEATHERPROOFING, TRACE HEATING, COOLING AND MECHANICAL PROTECTION*

10.20.01 Instrument air pipework
10.20.02 Process connections and assemblies, including fittings
10.20.03 Impulse pipework
10.20.04 Tubing and multi-core tubing
10.20.05 Manifolds
10.20.06 Instrument isolating valves
10.20.07 Drain, charging and other valves

10.21 TESTING AND COMMISSIONING: *SITE TESTING, PRE-COMMISSIONING AND COMMISSIONING COMPONENTS AND SYSTEMS AS REQUIRED WHERE NOT ALLOWED FOR ELSEWHERE*

10.21.01 Component/system site testing
10.21.02 Calibration
10.21.03 Pre-commissioning
10.21.04 Commissioning

10.22 CONTROL VALVES: *IF NOT INCLUDED WITH PIPEWORK*

10.23 CIVIL ASPECTS: *IF CIVIL WORKS DO NOT COVER, ACCOUNT IS TO BE TAKEN OF PRECAST TRENCHES, EXCAVATION, SAND, TILES, MARKERS, BACKFILL, COVERS, CONDUITS, DUCTS, SEALING, WEATHER PROTECTION AND GENERAL BUILDER'S WORK*

CAPPER PIPE SERVICES LIMITED

CAPPER HOUSE, DITTON ROAD,
WIDNES, CHESHIRE, WA8 0PG
TELEPHONE 051-420-6520
FAX 051-423-3934

PLANT INSTALLATION,
PIPEWORK AND
STEELWORK FABRICATION
AND ERECTION,
PROJECT MANAGEMENT.

OFFICES AT:
TEESIDE AND
BUCKINGHAMSHIRE   A MEMBER OF NORWEST HOLST

Scomark Engineering Ltd
(Regd. Office):
Hartshorne Road.
Woodville.
Burton-on-Trent.
Staffs. DE11 7JF

The Scomark Organisation is ideally structured to offer a complete service of highly specialised shop fabrication to BS 5750 and site mechanical installation. A compendium bid for the complete contract will obviate the tedium of matching multi-quotations and eliminate questions of dispute between different contractors.

We have an impressive record of completed prestigeous contracts and we offer our reputation in quality and service to discerning clients who seek a professional approach.

Telephone: 0283-218222   Telex: 341986 SCOMRK G   Telefax: 0283-226468

11.00

Remote fabrication/ pre-assembly facilities

11.01 Land and site development
11.02 Civil engineering
11.03 Structures
11.04 Plant
11.05 Mechanical services
11.06 Electrical services
11.07 Instrumentation and controls
11.08 Insulation
11.09 Protective coatings
11.10 Engineering design
11.11 Project management and/ or project charges
11.12 General factors
11.13 Transport

Pipework including spools and steelwork can be prefabricated/pre-assembled, or plant units can be modularized and constructed offsite to reduce onsite construction time or obviate site labour shortages. A separate full cost may be required for this work, or alternatively the estimator may be required to estimate the difference in cost between prefabrication and construction onsite.

For the full cost the estimator is referred to the main sections of the checklist but should also give consideration to the following list of differences in cost for construction at remote fabrication facilities.

11.01 LAND AND SITE DEVELOPMENT

11.01.01 Cost of separate yard and facilities
11.01.02 Any further development required to yard and load-out facilities or access thereto

11.02 CIVIL ENGINEERING

11.02.01 Concrete bases
11.02.02 Steel supports

11.03 STRUCTURES

11.03.01 Additional steelwork required to modularize
11.03.02 Bracing and additional steelwork required for fabrication
11.03.03 Bolt-up sections

11.04 PLANT

11.04.01 Additional packing plates under plant

11.05 MECHANICAL SERVICES

11.05.01 Testing pipe in shorter sections
11.05.02 Sealing ends of pipe
11.05.03 Fabrication of spool pieces for site hook-up
11.05.04 Extra welds in hook-up

11.06 ELECTRICAL SERVICES

11.06.01 Extra testing
11.06.02 Additional junction boxes
11.06.03 Rolling up cables including additional waste in length
11.06.04 Additional jointing

11.07 INSTRUMENTATION AND CONTROLS

11.07.01 Extra testing
11.07.02 Additional marshalling boxes
11.07.03 Rolling up cables including additional waste in length
11.07.04 Additional jointing
11.07.05 Protection of sensitive instruments (or remove and refit)

11.08 INSULATION

11.08.01 Protection of exposed ends of insulation
11.08.02 Separate insulation over hook-up welds and spool pieces

11.09 PROTECTIVE COATINGS

11.09.01 Separate protective coating over hook-up welds

11.10 ENGINEERING DESIGN

11.10.01 Additional design costs for prefabrications or modules

11.11 PROJECT MANAGEMENT AND/OR PROJECT CHARGES: *ADDITIONAL COSTS OF HAVING STAFF AND FACILITIES AT BOTH THE WORKSITE AND FABRICATION YARD*

11.11.01 Supervision and management
11.11.02 Inspection
11.11.03 Offices and welfare facilities
11.11.04 Telephone, telex, telefax and copying machine costs

11.12 GENERAL FACTORS

11.12.01 Effects of any difference in labour cost levels at fabrication yard
11.12.02 Savings constructing in permanent yard and/or in covered facilities

11.13 TRANSPORT

11.13.01 Any double handling of materials and transport to fabrication yard
11.13.02 Cranage and load-out costs
11.13.03 Temporary load-out and sea or road-fastening steel
11.13.04 Sea and/or road transport
11.13.05 Heavy lift cranage and facilities at site
11.13.06 Survey, preparation of accesses/roads/jetties for movement of modules

Darchem
Contracting UK Limited

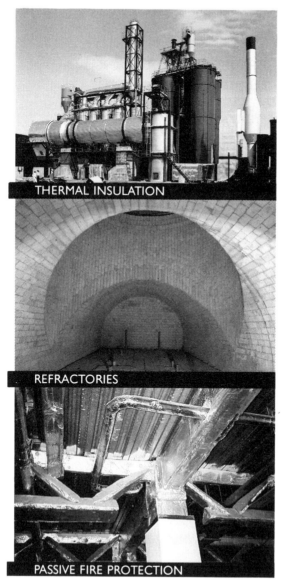

THERMAL INSULATION

REFRACTORIES

PASSIVE FIRE PROTECTION

Head Office
Faverdale Centre,
Faverdale Industrial Estate,
Darlington,
Co. Durham DL3 0QL
England

Telephone 0325 381301
Facsimile 0325 460619

Aberdeen
Telephone 0224 725509
Facsimile 0224 725516

Birmingham
Telephone 021 747 3731/6033
Facsimile 021 776 7066

Burnley
Telephone 0282 27467
Facsimile 0282 411711

Cardiff
Telephone 0222 371801
Facsimile 0222 221615

Glasgow
Telephone 041 445 2761
Facsimile 041 425 1151

Greenwich
Telephone 081 858 4851
Facsimile 081 853 4773

Sheffield
Telephone 0742 693711
Facsimile 0742 698550

Southampton
Telephone 0489 6771
Facsimile 0489 6402

12.00

Insulation

12.01 Type of insulation
12.02 Heat conservation, frost and personnel protection
12.03 Cold conservation and condensation protection
12.04 Cladding or surface finish

12.01 TYPE OF INSULATION

12.01.01 Fibre type
12.01.02 Moulded and formed insulation
12.01.03 Glass fibre
12.01.04 Mineral wool/cork
12.01.05 Polystyrene
12.01.06 Felt
12.01.07 Other specified insulation

12.02 HEAT CONSERVATION, FROST AND PERSONNEL PROTECTION

12.02.01 Boiler equipment, flues and ducts
12.02.02 Heat recovery plant
12.02.03 Other plant, equipment and machinery
12.02.04 Columns
12.02.05 Vessels
12.02.06 Drums
12.02.07 Tanks
12.02.08 Pumps
12.02.09 Pipes
12.02.10 Ducts

12.02.11 Buildings
12.02.12 Voids

12.03 COLD CONSERVATION AND CONDENSATION PROTECTION

12.03.01 Refrigeration plant
12.03.02 Cold rooms
12.03.03 Other plant, equipment and machinery
12.03.04 Pipes

12.04 CLADDING OR SURFACE FINISH

12.04.01 Metal cladding to any of the foregoing
12.04.02 Fibreglass finish to walking surfaces
12.04.03 Other surface finish

13.00

Protective coatings

13.01 General factors affecting cost
13.02 Fire protection
13.03 Below-ground protection
13.04 General painting

13.01 GENERAL FACTORS AFFECTING COST

13.01.01 Special scaffolding
13.01.02 Surface preparation
13.01.03 Materials, application and finish
13.01.04 Type of paint including oil, epoxy, cement based, chemical resistant, fire retardant, etc.
13.01.05 Touching up prepainted items after erection

13.02 FIRE PROTECTION: *MAKE ALLOWANCE FOR PAINT FINISH, CAST CONCRETE, SPRAYED FINISH, CLADDING OR SUCH OTHER FINISH AS MAY BE REQUIRED*

13.02.01 Surface preparation and any priming and painting below finish
13.02.02 Finish to buildings: walls
13.02.03 Finish to buildings: soffits
13.02.04 Finish to buildings: structural frame
13.02.05 Finish to steel structures and supports
13.02.06 Finish to plant, equipment machinery and vessels

13.03 BELOW-GROUND PROTECTION: *MAKE ALLOWANCE FOR PAINT FINISH, TAPE WRAPPING, ENCASING OR OTHER PROTECTION AS MAY BE REQUIRED*

13.03.01 Underground pipework and/or joints
13.03.02 Steel structural items or piling
13.03.03 Jetties or other structures below water level
13.03.04 Waterproofing

13.04 GENERAL PAINTING

13.04.01 Buildings (see also section 05.00)
13.04.02 Tanks, columns, vessels and similar
13.04.03 Plant, equipment, pumps and motors
13.04.04 Steel structures including enclosing housings, platforms, pipe racks, pipe supports, stairs, ladders and handrails
13.04.05 Pipework and supports
13.04.06 Electrical and instrumentation equipment
13.04.07 Heating and ventilation ducting and equipment
13.04.08 External facilities including fencing and gates
13.04.09 Finish below insulated pipes and surfaces
13.04.10 Motifs, colour coding, bands, lettering and panels

14.00

Offsite facilities

14.01 Land and site development
14.02 Civil engineering
14.03 Buildings or shelters
14.04 Structures
14.05 Plant
14.06 Mechanical services
14.07 Electrical services
14.08 Instrumentation and controls
14.09 Remote fabrication facilities
14.10 Insulation
14.11 Protective coatings

On process and certain other projects, work beyond the plot or battery limits of the new plant is usually referred to as the offsite facilities. The offsite facilities often include the import/export loading facilities which may be within the separate section 16.00.

Some offsite items are normally included in section 03.00 Land and site development, or section 05.06 External works to the building. A check should be made to ensure that items are not duplicated or missed. Each item is to include the civil engineering, insulation, painting, services and other associated items.

14.01 LAND AND SITE DEVELOPMENT

14.01.01 Road, rail or water access to the site
14.01.02 Services to and from the site
14.01.03 Refer to section 03.00 for other items

14.02 CIVIL ENGINEERING *(SEE SECTION 04.00 FOR A CHECKLIST OF ITEMS)*

14.02.01 Oversite excavation and filling
14.02.02 Piling

14.02.03 Foundations
14.02.04 Pits, basins and channels
14.02.05 Concrete ducts
14.02.06 Roads and pavings
14.02.07 Reinforcement and sundries
14.02.08 Drainage
14.02.09 Landscaping

14.03 BUILDINGS OR SHELTERS *(SEE SECTION 05.00 FOR A CHECKLIST OF ITEMS)*

14.03.01 Boiler houses
14.03.02 Compressor houses
14.03.03 Plant buildings
14.03.04 Workshops
14.03.05 Metering buildings
14.03.06 Analyser buildings
14.03.07 Chemical stores
14.03.08 Pump houses
14.03.09 Fire station
14.03.10 Medical centre/ambulance station
14.03.11 Substations/switch houses
14.03.12 Control rooms
14.03.13 Canteen
14.03.14 Rest rooms, changing rooms and welfare buildings
14.03.15 Administration buildings
14.03.16 Pit-head buildings
14.03.17 Other buildings

14.04 STRUCTURES *(SEE SECTION 06.00 FOR A CHECKLIST OF ITEMS)*

14.05 PLANT

14.05.01 Storage tanks or vessels (see section 07.00 for other items)

14.06 MECHANICAL SERVICES

14.06.01 Pipework
- (a) extending services, utilities and feedstock to plant
- (b) product and byproduct from plant
- (c) firewater lines (see section 08.00 for other items)

14.07 ELECTRICAL SERVICES

14.07.01 Extending supplies to plant
14.07.02 Supplies to all offsite plant and equipment
14.07.03 Supplies to all offsite buildings and where required to structures
14.07.04 Road and area lighting
14.07.05 Earthing
14.07.06 Lightning protection
14.07.07 Transformer compounds (see section 09.00 for other items)

14.08 INSTRUMENTATION AND CONTROLS

14.08.01 Control panels
14.08.02 Cable from panels to plant
14.08.03 Instruments to offsite plant and equipment and cabling back to panels (see section 10.00 for other items)

14.09 REMOTE FABRICATION FACILITIES *(SEE SECTION 11.00 FOR A CHECKLIST OF ITEMS IF APPLICABLE)*

14.10 INSULATION: *HEAT CONSERVATION, FROST AND PERSONNEL PROTECTION, COLD CONSERVATION AND CONDENSATION PROTECTION TO THE FOLLOWING*

14.10.01 Pipework
14.10.02 Plant, vessels and equipment (see section 12.00 for other items)

14.11 PROTECTIVE COATINGS: *FIRE PROTECTION, BELOW-GROUND PROTECTION AND PAINTING TO THE FOLLOWING*

14.11.01 Buildings
14.11.02 Tanks, columns, vessels, plant and equipment
14.11.03 Steel structures
14.11.04 Pipework and supports (see section 13.00 for other items)

15.00

Import/export pipelines

15.01 General factors affecting cost
15.02 Material cost
15.03 Easement cost
15.04 Right-of-way
15.05 Stringing
15.06 Trenching
15.07 Pipelaying
15.08 Crossings including additional reinstatement costs
15.09 Testing and commissioning
15.10 Reinstatement
15.11 Installations
15.12 Installation compounds
15.13 Telemetering and control systems
15.14 Intake and outfall structures and pump houses

15.01 GENERAL FACTORS AFFECTING COST

15.01.01 Fluid transported and working pressure
15.01.02 Environmental problems
15.01.03 Terrain: waterlogged or peat
15.01.04 Terrain: mountain
15.01.05 Terrain: woodland
15.01.06 Terrain: built-up areas

15.02 MATERIAL COST

15.02.01 Pipe
15.02.02 Thick wall pipe
15.02.03 Sleeves

15.02.04 Fittings including bends and all fittings required for block valve pigging and terminal facilities
15.02.05 Weight coating or anchors
15.02.06 Outfall nozzles

15.03 EASEMENT COST

15.03.01 Type of farmland and easement payment
15.03.02 Annual easement payment
15.03.03 Loss of crop
15.03.04 Accesses, pipe storage and fabrication areas

15.04 RIGHT-OF-WAY

15.04.01 Clearance of right-of-way
15.04.02 Temporary fencing
15.04.03 Topsoil stripping
15.04.04 Temporary roads in bad ground

15.05 STRINGING

15.05.01 Preparation of pipe dumps
15.05.02 Offloading in pipe dumps
15.05.03 Loading and stringing including skids
15.05.04 Cold bending
15.05.05 Provision of rollers, winches, buoyancy for offshore lines

15.06 TRENCHING

15.06.01 General trenching and backfilling
15.06.02 Trenching in bad ground or running sand
15.06.03 Dewatering
15.06.04 Trenching in rock
15.06.05 Trenching in built up areas
15.06.06 Underwater dredging, trenching, backfilling or protection
15.06.07 Sand padding
15.06.08 Imported trench backfill

15.07 PIPELAYING

15.07.01 Welding or jointing
15.07.02 Non-destructive testing
15.07.03 Wrapping
15.07.04 Holiday detection
15.07.05 Lower and lay
15.07.06 Cathodic protection including anodes
15.07.07 Weight coating or anchors
15.07.08 Laying of offshore live including all marine facilities, anchors and winches
15.07.09 Tie-in between offshore and onshore sections

15.08 CROSSINGS INCLUDING ADDITIONAL REINSTATEMENT COSTS

15.08.01 Thrust, bored or directional drilled crossings
15.08.02 Open cut: roads
15.08.03 Open cut: tracks
15.08.04 Open cut: railways
15.08.05 Open cut: water crossings
15.08.06 Open cut: ditches
15.08.07 Open cut: major services
15.08.08 Crossing other services, cattle crossings or other locations necessitating tie-ins
15.08.09 Provision for future motorways, road or river diversions, future ditch deepening or future land drainage systems

15.09 TESTING AND COMMISSIONING

15.09.01 Swabbing pig runs
15.09.02 Gauging pig runs
15.09.03 Sectional air testing
15.09.04 Provision and disposal of water
15.09.05 Water testing
15.09.06 Swabbing pig runs
15.09.07 Tie-in of test points
15.09.08 Drying

15.09.09 Gauging and/or intelligent pig runs
15.09.10 Testing material

15.10 REINSTATEMENT

15.10.01 Removal of temporary roads, skids, surplus soil and general site clearance
15.10.02 Ripping or loosening subsoil
15.10.03 Making good land drainage
15.10.04 Topsoil replacement
15.10.05 Rotovating
15.10.06 Stone picking
15.10.07 Re-seeding
15.10.08 Repairs to cross fences, walls and hedges
15.10.09 Removal (or leaving in) temporary fencing
15.10.10 Permanent road reinstatement
15.10.11 Track reinstatement
15.10.12 Marker posts

15.11 INSTALLATIONS

15.11.01 Intermediate block valve installations
15.11.02 Intermediate pigging installations
15.11.03 Terminal facilities including valves and pigging facilities
15.11.04 Metering and meter-proving facilities
15.11.05 Heaters
15.11.06 Compressor stations
15.11.07 Storage facilities, intermediate and at terminals

15.12 INSTALLATION COMPOUNDS

15.12.01 Purchase/rental of land
15.12.02 Access roads
15.12.03 Compound fencing and gates
15.12.04 Compound hard surfacing
15.12.05 Civil engineering work generally
15.12.06 Mechanical work

15.12.07   Electrical and instrumentation work
15.12.08   Tree planting in screening

15.13   TELEMETERING AND CONTROL SYSTEMS

15.14   INTAKE AND OUTFALL STRUCTURES AND PUMP HOUSES

15.14.01   Intake structure
15.14.02   Outfall structure
15.14.03   Pump house

SPON'S CONSTRUCTION COST AND PRICE INDICES HANDBOOK

B.A. Tysoe and M.C. Fleming

This unique handbook collects together a comprehensive and up-to-date range of indices measuring construction costs and prices. The authors give guidance on the use of the data making this an essential aid to accurate estimating.

Contents: Part A – Construction indices: uses and methodology. Uses of Construction Indices. Problems and methods of measurements. **Part B – Currently compiled construction indices.** Introduction. Output price indices. Tender price indices. DOE public sector building, 1968. DOE QSSD Index of Building tender prices. BCIS tender price index, 1974. DOE road construction tender price index, 1970. DOE price index for public sector house building (PIPSH), 1964. SLD housing tender price index (HTPI), 1970. DB&E tender price index, 1966. Cost indices. BCIS general building cost index, 1971. Spon's cost indices. Building cost index 1965. Electrical services cost index 1965. Civil engineering cost index 1970. Landscaping cost index 1976. APSAB cost index 1970. Building housing costs index 1973. SDD housing costs index 1970. BIA/BCIS house rebuilding costs index 1978. Association of Cost Engineers errected process plant indices 1958. BMCIS maintenance cost 1970. Summary comparison of indices and commentary. **Part C – Historical construction indices.** Introduction. Historical Cost and Price Indices. Maiwald's indices 1845–1938. Jones/Saville index 1845–1956. Venning index 1914. MOW/DOE 'CNC' indices 1939, 1946–1980 Q1. BRS measured work index 1939–1969 Q2. Summary comparison of indices and commentary. Appendix. General indices of prices. Index of total home cost. The retail price index. Index of capital goods cost. Glossary of Relevant Terms. Subject index.

April 1991   Hardback   224 pages   0 419 15330 6   c.£25.00

E & FN SPON
An Imprint of Chapman & Hall

16.00

Import/export loading facilities

16.01 Weighing devices, weighbridges and associated ticket systems
16.02 Flow metering facilities
16.03 Jetties and mooring facilities
16.04 Cranes
16.05 Fork-lift trucks and specialist vehicles/ attachments
16.06 Rail spurs, loops and railway systems
16.07 Special rail truck handling facilities
16.08 Bulk rail unloading facilities
16.09 Unloading and loading arms including screw unloaders
16.10 Barge tipplers
16.11 Conveyors, chutes and silos including magnetic separators, gauging, etc.
16.12 Vacuum transportation plants
16.13 Slurry ponds and equipment
16.14 Can, drum, bottling, packet filling, emptying and handling devices
16.15 Full and empty storage facilities
16.16 Road and parking facilities
16.17 Lifts, gantries, booms and special structures
16.18 Boom stackers, bucket wheel store/reclaim machines
16.19 Control systems
16.20 Batching plants

Support Services to Industry

CCG Contracting offer a wide range of on-site support services designed for industry. These may include the provision of site accommodation, warehousing, security, catering, cleaning and janitorial, medical care and construction village management.

What's more you can use these services singly, or as a complete package. Our involvement can be matched to meet your exact requirements.

For further details please contact **CCG Contracting** on **0786 834060**

· C O N T R A C T I N G ·

Steuart Road, Bridge of Allan, Stirling FK9 4JG

17.00

Operational and general services

17.01 Office machinery and paper handling
17.02 Communications
17.03 Transport
17.04 Maintenance facilities
17.05 Lighting
17.06 Safety and welfare
17.07 Fire services
17.08 Security
17.09 Weighbridge and similar provisions

17.01 OFFICE MACHINERY AND PAPER HANDLING

17.01.01 Collection/reception
    (a) sorting
    (b) distribution
    (c) franking
    (d) machines for franking, weighing, slitting and sealing
17.01.02 Telex machines, exchange lines and consumables
17.01.03 Facsimile machines, exchange lines and consumables
17.01.04 Copiers
    (a) A4 and A3 size
    (b) sorters
    (c) drawing reproduction machines
    (d) consumables
17.01.05 General office machinery
    (a) typewriters
    (b) word processors
    (c) drawing boards and drafting machines

            (d)   consumables
17.01.06 Computers
            (a)   main frames
            (b)   PC computers
            (c)   computer-aided design ancillaries such as printers, plotters, monitors and back-up
            (d)   consumables

17.02 COMMUNICATIONS

17.02.01 Telephones
            (a)   exchange and operators equipment
            (b)   internal lines and dedicated cabling
            (c)   external lines
            (d)   instruments
            (e)   answering machines
            (f)   monitors and loggers
17.02.02 Radio
            (a)   fixed transmitters/receivers
            (b)   portable transmitters/receivers
            (c)   car and personal radio telephones
            (d)   staff location and bleeper systems
17.02.03 Public address
            (a)   amplifier
            (b)   input stations
            (c)   output stations
            (d)   dedicated cabling
17.02.04 Video
            (a)   closed circuit TV systems
            (b)   recorders

17.03 TRANSPORT

17.03.01 Barges, tugs, etc.
17.03.02 Shunting locomotives and rolling stock
17.03.03 Tankers, lorries
17.03.04 Cars, vans
17.03.05 Personnel transport
17.03.06 Motor cycles, cycles

17.03.07　Lift and hoist trucks
17.03.08　Mobile trucks and platforms
17.03.09　Aircraft and helicopters
17.03.10　Associated special maintenance facilities

17.04 MAINTENANCE FACILITIES

17.04.01　Site maintenance including landscape and gardening
17.04.02　Building maintenance
17.04.03　Machinery and equipment maintenance
17.04.04　Instrument maintenance
17.04.05　Electrical maintenance including lifts
17.04.06　Vehicle and aircraft maintenance
17.04.07　Cleaning facilities including vacuum cleaning and windows
17.04.08　Offloading, handling and lifting facilities
17.04.09　Cranes and portable hoists
17.04.10　Scaffolds, ladders, trestles, etc.
17.04.11　Portable equipment (e.g. concrete mixers, pumps, air compressors, welding sets, small hand tools, etc.)
17.04.12　Associated special safety gear (e.g. goggles, welding aprons, spark-proof tools, low voltage tools)
17.04.13　Spares

17.05 LIGHTING

17.05.01　Emergency lighting
17.05.02　Portable lighting
17.05.03　Warning lights
17.05.04　Traffic lights
17.05.05　Generators

17.06 SAFETY AND WELFARE

17.06.01　Drinking water points
17.06.02　Showers and baths
17.06.03　Emergency showers and baths
17.06.04　Eyewash and first-aid kits
17.06.05　Ambulances

17.06.06   Protective clothing
17.06.07   Portable equipment (resuscitators, stretchers, etc.)
17.06.08   Breathing sets, lifebelts (e.g. for tank entry)
17.06.09   Smoking areas
17.06.10   First-aid stations and equipment

17.07   FIRE SERVICES

17.07.01   Firemains and hydrants
17.07.02   Fire appliances (engines, tenders, pumps, hoses)
17.07.03   Foam and CO_2 systems
17.07.04   Fire alarm systems
17.07.05   Personnel clothing and equipment
17.07.06   Safety gear (breathing equipment, ropes, etc.)

17.08   SECURITY

17.08.01   Security fencing
17.08.02   Time clocks
17.08.03   Watchmen
17.08.04   Intruder alarms
17.08.05   Home Office, Customs and Excise and similar provisions
17.08.06   Security patrols

17.09   WEIGHBRIDGE AND SIMILAR PROVISIONS

17.09.01   Weighbridge and similar provisions

18.00

Temporary and/or common site facilities

18.01 Utilities including distribution and outlets and, where appropriate, charges
18.02 Office accommodation
18.03 Safety and welfare buildings
18.04 Stores and workshops
18.05 Contractors' areas
18.06 Parking and personnel transport
18.07 Site security
18.08 Hardstandings
18.09 Scaffolding
18.10 General site transport, cranage and plant
18.11 Contractors providing common services to site
18.12 Rates

The estimator should ascertain the scope of temporary facilities required for the contractor and any common services to be provided onsite for use by other contractors or the client. Allowance should be made for provision, operation, maintenance and termination of the facilities.

18.01 UTILITIES INCLUDING DISTRIBUTION AND OUTLETS AND, WHERE APPROPRIATE, CHARGES

18.01.01 Electricity
18.01.02 Water
18.01.03 Sewage and drainage
18.01.04 Compressed air and gases
18.01.05 Telephone/telex/telefax
18.01.06 Maintenance

18.02 OFFICE ACCOMMODATION

18.02.01 Offices and toilets
18.02.02 Car parking
18.02.03 Cleaning and maintenance
18.02.04 Jobsite signboard

18.03 SAFETY AND WELFARE BUILDINGS

18.03.01 Rest and changing rooms
18.03.02 Toilets and washrooms
18.03.03 Lockers and drying rooms
18.03.04 Canteen (including vending machines about site)
18.03.05 Medical suite/first-aid post
18.03.06 Ambulance
18.03.07 Medical attendant
18.03.08 Smoking areas
18.03.09 Cleaning and maintenance

18.04 STORES AND WORKSHOPS

18.04.01 Compound including preparation and fencing
18.04.02 Stores building
18.04.03 Stores cranage and equipment
18.04.04 Stores labour
18.04.05 Workshops
18.04.06 Cleaning and maintenance

18.05 CONTRACTORS' AREAS

18.05.01 Preparation
18.05.02 Fencing
18.05.03 Access to areas

18.06 PARKING AND PERSONNEL TRANSPORT

18.06.01 Parking areas for workforce and staff and parking control
18.06.02 Personnel transport offsite
18.06.03 Personnel transport onsite
18.06.04 Maintenance

18.07 SITE SECURITY

18.07.01 Identification pass system
18.07.02 Operation of time clocking
18.07.03 Site guards including guardhouse
18.07.04 Security floodlighting
18.07.05 Microwave and infrared systems

18.08 HARDSTANDINGS

18.08.01 Hardstanding to working areas and access ways

18.09 SCAFFOLDING

18.09.01 Scaffolding and platforms
18.09.02 Lifts and hoists

18.10 GENERAL SITE TRANSPORT, CRANAGE AND PLANT

18.10.01 General site transport
18.10.02 General excavators/loaders
18.10.03 Cranage including tower cranes
18.10.04 Cement storage and mixing facilities and batch plants
18.10.05 Refuelling facilities
18.10.06 Floodlighting

18.11 CONTRACTORS PROVIDING COMMON SERVICES TO SITE

18.11.01 Scaffolding
18.11.02 Cleaning
18.11.03 Catering
18.11.04 Painting
18.11.05 Office management and services
18.11.06 Cranes

18.12 RATES

18.12.01 Local taxation on buildings

::: DEBORAH GRAYSTON SCAFFOLDING

A Complete and Professional Service to Industry

For all your access requirements throughout the
Construction – Module Fabrication – Maintenance
Building and Civil Engineering Industries we provide
the following:

**DESIGN • MANAGEMENT • SAFETY AND TRAINING
MATERIAL • LABOUR • INSPECTION**

NATIONAL SAFETY AWARD
achieved **NINE** Years in Succession

NECEA Safe Working Award achieved on first application

Deborah Grayston Scaffolding Ltd
10 South Parade, Wakefield, West Yorkshire WF1 1LS
Tel: Wakefield (0924) 378222 Telex: 556227 DEB G
Fax: (0924) 366250

A BET PLANT SERVICES COMPANY

19.00

Infrastructure

19.01 Facilities during construction

19.02 Facilities left after construction

19.01 FACILITIES DURING CONSTRUCTION

19.01.01 Housing encampments
19.01.02 Medical facilities
19.01.03 Welfare and recreational facilities
19.01.04 Schooling
19.01.05 Roads and site layout
19.01.06 Harbour facilities
19.01.07 Airstrips and helipads
19.01.08 Fencing
19.01.09 Sewage systems
19.01.10 Utilities
19.01.11 Staffing of facilities

19.02 FACILITIES LEFT AFTER CONSTRUCTION

19.02.01 Refurbishment and handover of any of the above
19.02.02 Other contributions to local or national community

Redpath Engineering Services Ltd.

Portrack Lane
Stockton-on-Tees
Cleveland TS18 2PR
Tel   : (0642) 673333
Telefax: (0642) 671033
Telex: 58210 THRES G

Chemical, Pharmaceutical, Offshore, Nuclear, Commercial and General Industrial Sectors in:
* Pipework design, fabrication and erection
* Mechanical plant installation
* Electrical and instrumentation installation
* Modularised skid mounted units
* General Engineering fabrication
* Pressure Vessel manufacture
* Engineering personnel on a contract hire basis

20.00

Modifications and alterations to existing plant

20.01 Demolition or removal
20.02 New work
20.03 Alterations, refurbishment and updating
20.04 Maintenance

20.01 DEMOLITION OR REMOVAL

20.01.01 Gas free/decommission before commencement
20.01.02 Asbestos removal
20.01.03 Plant and equipment
20.01.04 Pipework
20.01.05 Electrical, instrumentation and telecommunications cabling and apparatus
20.01.06 Structures
20.01.07 Buildings
20.01.08 Underground foundations, services and paving layouts

20.02 NEW WORK (SEE ALL SECTIONS OF THIS CHECKLIST)

20.03 ALTERATIONS, REFURBISHMENT AND UPDATING

20.03.01 Gas free/decommission before commencement
20.03.02 Remedial work after asbestos removal
20.03.03 Plant and equipment
20.03.04 Pipework
20.03.05 Electrical, instrumentation and telecommunications cabling and apparatus
20.03.06 Structures
20.03.07 Buildings
20.03.08 Temporary facilities to keep plant operational
20.03.09 Hoardings, barriers, dust protection, etc.

20.04 MAINTENANCE

20.04.01 Maintenance painting
20.04.02 Other general maintenance work

21.00

Spares

21.01 Capital spares
21.02 Consumable spares
21.03 Evection spares

21.04 Spares for potential obsolescent equipment

SPON'S
Architects' and Builders' Price Book 1991

116th Annual Edition

Edited by *Davis Langdon & Everest*

With labour rates increasing, rising materials prices and competition hotting up, dependable *market based* cost data is essential for successful estimating and tendering. Compiled by the world's largest quantity surveyors, Spon's are the only price books geared to building tenders and the *market conditions* that affect building prices.

Hardback   over 900 pages

0 419 16790 0   £49.50

SPON'S
Mechanical and Electrical Services Price Book 1991

22nd Annual Edition

Edited by *Davis Langdon & Everest*

Now better than ever, all of Spon's M & E prices have been reviewed in line with current tender values, providing a unique source of market based pricing information. The Approximate Estimating Section has been greatly expanded giving elemental rates for four types of development: computer data centres, hospitals, hotels and factories. The only price book dedicated exclusively to mechanical and electrical services.

Hardback   over 730 pages

0 419 16800 1   £52.50

E & FN SPON
An Imprint of Chapman & Hall

22.00

Engineering design and procurement

22.01 Proposal engineering
22.02 Project management
22.03 Process engineering
22.04 Civil and structural engineering
22.05 Vessel and heat transfer engineering
22.06 Machinery and mechanical engineering
22.07 Electrical engineering
22.08 Instrumentation engineering
22.09 Piping engineering
22.10 Safety engineering
22.11 Quality assurance
22.12 Quality control
22.13 General engineering activities
22.14 General project costs

Additional details for section 22.00 are included in Appendix 1, to which reference should be made in conjunction with the above main list; see section 22.13 for details of general activities.

22.01 PROPOSAL ENGINEERING

22.01.01 Proposal co-ordination
22.01.02 Estimating
22.01.03 Planning
22.01.04 Contract and legal matters
22.01.05 Technology input
22.01.06 Secretarial and clerical work

22.02 PROJECT MANAGEMENT

22.02.01 Project managers
22.02.02 Project engineering
22.02.03 Project controls and accounting
22.02.04 Procurement and expediting
22.02.05 Contracts
22.02.06 Consultants and specialists
22.02.07 Secretarial and clerical work
22.02.08 Documentation

22.03 PROCESS ENGINEERING

22.03.01 Basic process design
22.03.02 Flowsheets
22.03.03 Schedules and lists
22.03.04 Safety considerations
22.03.05 Commissioning
22.03.06 General activities

22.04 CIVIL AND STRUCTURAL ENGINEERING

22.04.01 Architectural design
22.04.02 Surveys
22.04.03 Site development
22.04.04 Underground services
22.04.05 Foundations
22.04.06 Structural concrete
22.04.07 Steelwork
22.04.08 Documentation
22.04.09 General activities

22.05 VESSEL AND HEAT TRANSFER ENGINEERING

22.05.01 Basic design: storage tanks
22.05.02 Detailed design: storage tanks

22.05.03 Basic design: pressure vessels and agitators
22.05.04 Detailed design: pressure vessels and agitators
22.05.05 Basic and detailed design: columns and stacks
22.05.06 Basic and detailed design: heat transfer
22.05.07 Fired heaters
22.05.08 Disintegration equipment
22.05.09 General activities

22.06 MACHINERY AND MECHANICAL ENGINEERING

22.06.01 Rotating machinery design
22.06.02 Mechanical handling design
22.06.03 Mechanical packages design
22.06.04 Services
22.06.05 General activities

22.07 ELECTRICAL ENGINEERING

22.07.01 Specifications
22.07.02 Requisition and data sheets
22.07.03 Diagrams
22.07.04 Layouts
22.07.05 Schedules and schematics
22.07.06 Detail drawings
22.07.07 Manuals
22.07.08 Safety
22.07.09 General activities

22.08 INSTRUMENT ENGINEERING

22.08.01 Conceptual design
22.08.02 Layouts
22.08.03 Hook-ups
22.08.04 Indexes and schedules
22.08.05 Schematic diagrams
22.08.06 Panel design
22.08.07 Instrument specifications

22.08.08　System specifications
22.08.09　Material take-off
22.08.10　Process instrument specifications
22.08.11　Engineering line diagrams
22.08.12　Communicating equipment design
22.08.13　Packaged equipment interfaces
22.08.14　General activities

22.09　PIPING ENGINEERING

22.09.01　Materials engineering
22.09.02　Schedules
22.09.03　Piping models: notional
22.09.04　Piping models: detail
22.09.05　Piping studies
22.09.06　Layouts: prepared manually
22.09.07　Layouts: computer designed
22.09.08　General arrangement drawings: prepared manually
22.09.09　General arrangement drawings: computer designed
22.09.10　Isometrics: prepared manually
22.09.11　Isometrics: computer designed
22.09.12　Piping stress engineering
22.09.13　Material take-off: prepared manually
22.09.14　Material take-off: by computer
22.09.15　Flow and line diagrams
22.09.16　Weight control
22.09.17　Insulation
22.09.18　General activities

22.10　SAFETY ENGINEERING

22.10.01　General safety
22.10.02　Safety criteria
22.10.03　Design safety
22.10.04　Firewater distribution
22.10.05　Safety and fire-fighting equipment
22.10.06　General activities

22.11 QUALITY ASSURANCE

22.11.01 Design phase
22.11.02 Procurement phase

22.12. QUALITY CONTROL

22.12.01 Vendor quality control

22.13 GENERAL ACTIVITIES

22.13.01 Procurement/expediting
22.13.02 Safety reviews
22.13.03 Approve client's drawings
22.13.04 Vendor print: control and approval
22.13.05 Site liaison
22.13.06 Local authority liaison
22.13.07 Inter-departmental meetings
22.13.08 Site meetings and visits
22.13.09 Requisitions
22.13.10 Bid analysis sheets
22.13.11 As-built drawing preparation
22.13.12 Construction support
22.13.13 Weight reports
22.13.14 Final documentation
22.13.15 Project close-out reports
22.13.16 Supervision

22.14 GENERAL PROJECT COSTS

22.14.01 Project office costs and expenses
22.14.02 Records and communications
22.14.03 Staff costs
22.14.04 Fees and charges
22.14.05 Finance
22.14.06 Other requirements and services
22.14.07 Consignment costs

Maximise your benefits with

CURRIE & BROWN
Quantity Surveying Cost & Contract Engineering Project Management

in the Financial Control of your Project

Worldwide Experience Includes:

Buildings, Hotels, Offices, Leisure Developments, Mechanical & Electrical Installations, Process, Petrochemical, Oil & Gas Plants, Offshore Installations, Airports, Marine Terminals, Mining, Power Stations, etc.

Work Undertaken:

Estimates, Contract Documentation, Quantity Surveying, Variations, Claims, Cost Control and Administration.

Currie & Brown,
The Red House, High Street,
Redbourn, St. Albans,
Herts AL3 7LE
Telephone: 0582 793003
Telex: 825491 Currie G
Fax: 0582 793000

Initially telephone or fax a brief outline of your requirements to our office in St. Albans, UK. We will then arrange our local office/associated company to respond to your enquiry.

Principal Offices & Associations:
18 Offices in the UK. Also in Australia, New Zealand, Indonesia, Japan, U.S.A., Malaysia, Singapore, France, Italy, Netherlands, Norway and Spain.

PROVIDING CONSTRUCTIVE SOLUTIONS

23.00

Project construction management and project charges

23.01 Head office project management: construction phase
23.02 Finalization of design
23.03 Procurement
23.04 Subcontracts
23.05 Site management and supervision
23.06 Site support and services
23.07 Staff costs
23.08 Site facilities
23.09 Records
23.10 Overheads

23.01 HEAD OFFICE PROJECT MANAGEMENT: CONSTRUCTION PHASE

23.01.01 Engineering departmental liaison
23.01.02 Site liaison
23.01.03 Vendor liaison
23.01.04 Construction procurement
23.01.05 Freight and delivery direction
23.01.06 Expediting
23.01.07 Continuing head office management services

23.02 FINALIZATION OF DESIGN

23.02.01 Tails engineering (all technologies)
23.02.02 Design work required to develop contract

23.03 PROCUREMENT

23.03.01 Complete design phase
23.03.02 Requisition and procurement-site material requirements
23.03.03 Complete expediting

23.04 SUBCONTRACTS

23.04.01 Complete preparation and negotiation of subcontracts
23.04.02 Administer contract and subcontracts
23.04.03 Agreement of variations, extra works, claims, progress
23.04.04 Payments and final account settlements with subcontractors, suppliers and client

23.05 SITE MANAGEMENT AND SUPERVISION

23.05.01 Construction manager
23.05.02 Construction engineers
23.05.03 Resident site manager
23.05.04 Cost engineers/quantity surveyors
23.05.05 Planning engineers
23.05.06 Material controllers
23.05.07 Quality assurance and quality control engineers
23.05.08 Specialist engineers and consultants
23.05.09 Industrial relations officer
23.05.10 Safety officer
23.05.11 Welding engineer
23.05.12 Buyers
23.05.13 Site accountants
23.05.14 Subcontract administrators

23.06 SITE SUPPORT AND SERVICES

23.06.01 Secretarial services
23.06.02 Clerks
23.06.03 Cars/drivers
23.06.04 Fees of other consultants or agents

23.06.05　Licences and permits
23.06.06　Inspection and testing
23.06.07　Models and publicity

23.07　STAFF COSTS

23.07.01　Travel and subsistence expenses
23.07.02　Expenses for dependants (e.g. travel, accommodation and schooling)
23.07.03　Training of personnel (operations, maintenance and safety)

23.08　SITE FACILITIES: *(SEE ALSO SECTION 18.00 TEMPORARY FACILITIES)*

23.08.01　Utilities, office accommodation, safety and welfare, buildings, stores, workshops, contractors' areas, parking and personnel transport, site security, hardstandings, scaffolding, general site transport, cranage, plant and local authority rates

23.09　RECORDS

23.09.01　Final documentation (test and inspection records, operating and instruction manuals)
23.09.02　As-built drawings

23.10　OVERHEADS

23.10.01　Head office overheads
23.10.02　Financing
23.10.03　Escalation
23.10.04　Insurances
23.10.05　Bonds and guarantees
23.10.06　Profit

SPON'S
Civil Engineering and Highway Works
Price Book 1991

5th Annual Edition

Edited by *Davis Langdon & Everest*

"Unquestionably, this book will be required by all estimators involved in civil engineering works, quantity surveyors will also find it essential for their shelves." - *Civil Engineering Surveyor*

This is more than a price book, Spon's Civil Engineering and Highway Works is a comprehensive manual for all civil engineering estimators and quantity surveyors.

Hardback   over 890 pages

0 419 16810 9   £55.00

UPDATED EVERY 3 MONTHS FREE!

SPON'S
Landscape and External Works
Price Book 1991

11th Annual Edition

Edited by *Davis Langdon & Everest* and *Lovejoy*

This is an indispensible handbook for landscape architects, surveyors and architects. As the only price book of its kind, Spon's Landscape has become the industry's standard guide for compiling estimates, specifications, bills of quantity, and works schedules.

Hardback   over 250 pages

0 419 16820 6   £42.50

E & FN SPON
An Imprint of Chapman & Hall

24.00

Contract conditions

24.01 Design and scope changes
24.02 Guarantee of performance of plant
24.03 Performance bond
24.04 Delays, extension of time and liquidated damages
24.05 Adverse physical conditions
24.06 Excepted risks
24.07 Open-ended liabilities
24.08 Third party supply
24.09 Co-ordination with existing plant and others
24.10 *Force majeure*
24.11 Patent rights and royalties
24.12 Samples and tests
24.13 Site security
24.14 Insurances
24.15 Variations and additional work
24.16 Interim payment provisions and retention
24.17 Special computer security
24.18 Certification and completion requirements
24.19 As-constructed documentation
24.20 Maintenance
24.21 Arbitration

Contract conditions should be reviewed to ascertain potential liabilities for which no allowances have been made against other sections of this checklist. It is necessary to check which items are reimbursable and those for which allowances must be included here, i.e. any or all of the above. A check should be made for liabilities against each contract clause notwithstanding the list above (see also section 23.00 Project management and/or project charges, and section 25.00 Overseas projects).

NEW EDITION
STANDARD METHOD OF SPECIFYING FOR MINOR WORKS
3rd edition
L Gardiner

The purpose of this book is to improve communication between builder and client by outlining a method of specification united to minor works. In this edition further up-dating has been carried out mainly with reference to British Standard Specification or Codes of Practice, Forms of Contract, and the 'Guarantee" and the 'Warranty' schemes of the leading building organisations. It also introduces the requirements under the new Local Government & Housing Act 1989.

Contents: Acknowledgements. Author's preface. **Section 1:** Introduction and examples of schedules of works. **Section 2:** Preliminaries and general matters, materials, workmanship and usage of sundry building terms. **Section 3:** Schedules of work – presentation and order. **Section 4:** Contract. Index.

E. & F.N. Spon   March 1991   Hardback   208 pages   0 419 15520 1   £25.00

NEW
SPON'S BUDGET ESTIMATING HANDBOOK
Edited by **Spain and Partners**, Consulting Quantity Surveyors, Merseyside, UK

This new reference book from Spon is the first estimating guide to concentrate entirely on approximate estimating for quantity surveyors, architects, and developers. This handbook provides a one-stop reference point in preparing estimates, so saving valuable time and increasing the quality and quantity of approximate cost data.

Contents: Preface – explaining the contents and how to use them including assessments of accuracy limits. **Part One – Building:** Square metre prices. Elemental cost plans – for different types of buildings. Composite costs – all-in rates. **Part Two – Civil Engineering:** Approximate costs – using formulae, e.g. the cost of water treatment works using the daily throughput of water. Composite costs – all-in rates, e.g. pipelines including trench bedding, pipework and backfilling. **Part Three – Mechanical and Electrical Work:** Composite costs – all-in rates. **Part Four – Reclamation and Landscaping:** Composite costs – all-in rates. **Part Five – Land and Development:** Land values. Development costs. **Part Six – Fees:** Professional fees. Index.

E. & F.N. Spon   August 1990   Hardback   200 pages   0 419 14780 2   £34.50

ORDER FORM

_____ copy/ies of Spon's Budget Estimating Handbook
_____ copy/ies of Standard Method of Specifying for Minor Works (Pub: March 1991)
☐ I enclose a cheque/PO for £_____ payable to E. & F.N. Spon
Please debit my ☐ Visa   ☐ Access   ☐ AmEx   ☐ Diners Club
Card No. _____ Expiry date _____
Signature _____
Address _____
_____ Postcode _____
Please return your order to: E & FN SPON, 2-6 Boundary Row, London SE1 8HN

E & FN SPON
An Imprint of Chapman & Hall

25.00

Overseas projects

25.01 Design fabrication and construction
25.02 Consignment costs
25.03 Fees
25.04 Commissions
25.05 Taxes and dues
25.06 Local territory costs
25.07 Expatriate staff and labour
25.08 Additional costs
25.09 Other risks

25.01 DESIGN FABRICATION AND CONSTRUCTION

25.01.01 Compliance with local standards, specifications, practices and quality assurance

25.02 CONSIGNMENT COSTS

25.02.01 Export packing
25.02.02 Weather protection
25.02.03 Freight to and from ports
25.02.04 Loading and offloading (land and sea freight)
25.02.05 Shipment
25.02.06 Demurrage
25.02.07 Customs delays and other delays at ports
25.02.08 Storage
25.02.09 Double handling and re-handling
25.02.10 Insurance, loss, damage, etc.
25.02.11 Premiums for war zones

25.03 FEES

25.03.01 UK agents
25.03.02 Customs agents
25.03.03 Other agents and suppliers' agents
25.03.04 Translations
25.03.05 Legal

25.04 COMMISSIONS

25.04.01 Bank charges including letters of credit
25.04.02 Export credit guarantee
25.04.03 Buying
25.04.04 Consulate charges
25.04.05 Gratuities
25.04.06 Local agent

25.05 TAXES AND DUES

25.05.01 Import duties and port dues
25.05.02 Transmission taxes
25.05.03 Other local taxes

25.06 LOCAL TERRITORY COSTS

25.06.01 Legislation: use of local labour and materials
25.06.02 Quality, cost and availability of local labour, plant and materials
25.06.03 Compliance with local regulations
25.06.04 Local trade practices, demarcation, etc.
25.06.05 Availability and suitability of local roads, utilities and housing

25.07 EXPATRIATE STAFF AND LABOUR

25.07.01 Overseas allowances
25.07.02 Accommodation
25.07.03 Air fares and travel expenses
25.07.04 Leave
25.07.05 Local income tax
25.07.06 Work permits, visas and licences
25.07.07 Site welfare facilities
25.07.08 Schooling
25.07.09 Medical expenses
25.07.10 Local transport

25.08 ADDITIONAL COSTS

25.08.01 Higher level of initial spares
25.08.02 Higher level of construction, commissioning and operational spares
25.08.03 Extra cost of emergency freight

25.09 OTHER RISKS

25.09.01 Major economic dislocation
25.09.02 Payment: currency and place of payment
25.09.03 Risks not covered by any export credit guarantee
25.09.04 Settlement of disputes

SPON'S CONSTRUCTION COST AND PRICE INDICES HANDBOOK

B.A. Tysoe and M.C. Fleming

This unique handbook collects together a comprehensive and up-to-date range of indices measuring construction costs and prices. The authors give guidance on the use of the data making this an essential aid to accurate estimating.

Contents: Part A – Construction indices: uses and methodology. Uses of Construction Indices. Problems and methods of measurements. **Part B – Currently compiled construction indices.** Introduction. Output price indices. Tender price indices. DOE public sector building, 1968. DOE QSSD Index of Building tender prices. BCIS tender price index, 1974. DOE road construction tender price index, 1970. DOE price index for public sector house building (PIPSH), 1964. SLD housing tender price index (HTPI), 1970. DB&E tender price index, 1966. Cost indices. BCIS general building cost index, 1971. Spon's cost indices. Building cost index 1965. Electrical services cost index 1965. Civil engineering cost index 1970. Landscaping cost index 1976. APSAB cost index 1970. Building housing costs index 1973. SDD housing costs index 1970. BIA/BCIS house rebuilding costs index 1978. Association of Cost Engineers errected process plant indices 1958. BMCIS maintenance cost 1970. Summary comparison of indices and commentary. **Part C – Historical construction indices.** Introduction. Historical Cost and Price Indices. Maiwald's indices 1845–1938. Jones/Saville index 1845–1956. Venning index 1914. MOW/DOE 'CNC' indices 1939, 1946–1980 Q1. BRS measured work index 1939–1969 Q2. Summary comparison of indices and commentary. Appendix. General indices of prices. Index of total home cost. The retail price index. Index of capital goods cost. Glossary of Relevant Terms. Subject index.

April 1991   Hardback   224 pages   0 419 15330 6   c.£25.00

E & FN SPON
An Imprint of Chapman & Hall

26.00

Commissioning, initial operation and training

26.01 Commissioning
26.02 Initial operation
26.03 Training
26.04 Final clean-up and making good

26.01 COMMISSIONING

26.01.01 Ascertaining extent of work (i.e. assistance only to client)
26.01.02 Specialist engineers and operators
26.01.03 Commissioning equipment
26.01.04 Vendor's commissioning engineers and operators
26.01.05 Power supplies and consumables
26.01.06 Remedial work including special studies, investigations and redesigning as required

26.02 INITIAL OPERATION

26.02.01 Start-up costs
26.02.02 Operators and labour
26.02.03 Power supplies, consumables and raw materials
26.02.04 Working alongside and training client's operators

26.03 TRAINING

26.03.01 Training client's staff: classroom
26.03.02 Training client's staff: during operation (see also section 26.02.04)
26.03.03 Translations of manuals to local language

26.04 FINAL CLEAN-UP AND MAKING GOOD

ASSOCIATION OF COST ENGINEERS

IF YOU WANT TO KNOW MORE ABOUT MEMBERSHIP AND OTHER SERVICES AND PUBLICATIONS THE ASSOCIATION PROVIDES PLEASE TELEPHONE OR WRITE TO :-

Administrative Secretary
The Association of Cost Engineers
Lea House
5 Middlewich Road
Sandbach
Cheshire CW11 9XL
Tel : 0270 764798
Fax : 0270 766180

27.00

Provisional sums, contingencies and escalation

27.01 Provisional and prime cost (PC) sums to be included on direction of client
27.02 Provisional and prime cost (PC) sums included by estimator
27.03 Contingency sums to be included on direction of client
27.04 Contingency allowance included by estimator
27.05 Escalation and inflation

27.01 PROVISIONAL AND PRIME COST (PC) SUMS TO BE INCLUDED ON DIRECTION OF CLIENT

27.02 PROVISIONAL AND PRIME COST (PC) SUMS INCLUDED BY ESTIMATOR

27.03 CONTINGENCY SUMS TO BE INCLUDED ON DIRECTION OF CLIENT

27.04 CONTINGENCY ALLOWANCE INCLUDED BY ESTIMATOR WHICH MAY INCLUDE ANY OF THE FOLLOWING OTHER ITEMS

27.04.01 Inclement weather
27.04.02 Industrial relations problems
27.04.03 Delays by others
27.04.04 Historical growth factor which may include design development and construction changes, unforeseen items, additional work, variations and claims
27.04.05 Quality of information and data and estimate accuracy
27.04.06 Risk

27.05 ESCALATION AND INFLATION IN THE FOLLOWING

27.05.01 Engineering and design
27.05.02 Construction and labour
27.05.03 Construction plant
27.05.04 Construction materials
27.05.05 Site overheads and management
27.05.06 Fixed price element
27.05.07 Differences between actual increase and any reimbursement by an index
27.05.08 Market forces

Appendix

Engineering design and procurement additional details*

22.01 Proposal engineering
22.02 Project management
22.03 Process engineering
22.04 Civil and structural engineering
22.05 Vessel and heat transfer engineering
22.06 Machinery and mechanical engineering
22.07 Electrical engineering
22.08 Instrumentation engineering
22.09 Piping engineering
22.10 Safety engineering
22.11 Quality assurance
22.12 Quality control
22.13 General activities
22.14 General project costs

22.01 PROPOSAL ENGINEERING

22.01.01 Proposal co-ordination
 (a) pre-bid client negotiation
 (b) prepare basic data
 (c) department liaison
 (d) co-ordinate presentation to client
 (e) post-tender negotiations
22.01.02 Estimating
 (a) engineering manhour estimate
 (b) equipment and material estimate

*ELD = engineering line diagram; HVAC = heating, ventilating and air conditioning; ULD = utility line diagram.

    (c) construction estimate
    (d) commissioning estimate
    (e) project and construction estimates
22.01.03 Planning
    (a) programme for proposal
    (b) overall project programme
    (c) manhour histograms
    (d) 'S' curves for cashflow
22.01.04 Contract and legal matters
    (a) tender documents
    (b) advise on financial and cashflow requirements
    (c) preliminary contract documents
    (d) obtain ECGD and other insurance quotations
    (e) post-tender negotiations
22.01.05 Technology input
    (a) data for estimating
    (b) engineering manhour estimates
    (c) staffing programmes
    (d) material take-offs if required
    (e) input to tender documentation
22.01.06 Secretarial and clerical work
    (a) general assistance and word processing
    (b) collating and producing tender documents

22.02 PROJECT MANAGEMENT: *NO ADDITIONAL DETAILS ARE GIVEN FOR THIS ACTIVITY SINCE THE MAJORITY OF PERSONNEL EMPLOYED ON THE PROJECT TEAM ARE GENERALLY ENGAGED ON A PROGRAMME BASIS RATHER THAN A SCOPE BASIS; CARE SHOULD BE TAKEN TO ASCERTAIN THE MAGNITUDE OF THE PROJECT AND THE PROGRAMME PROPOSED, IN ORDER TO ASSESS THE PERSONNEL REQUIREMENTS IN EACH DISCIPLINE*

22.03 PROCESS ENGINEERING

22.03.01 Basic process design
    (a) heat and material balances
    (b) process studies

- (c) utility systems
- (d) process data sheets

22.03.02 Flowsheets
- (a) process flow diagrams
- (b) engineering flow diagrams
- (c) utility flow diagrams

22.03.03 Schedules and lists
- (a) equipment lists
- (b) line lists
- (c) controls and instrument schedules

22.03.04 Safety considerations
- (a) safety guidelines
- (b) safety review
- (c) safety reports

22.03.05 Commissioning
- (a) manuals and dossiers
- (b) pre-commissioning
- (c) commissioning

22.04 CIVIL AND STRUCTURAL ENGINEERING

22.04.01 Architectural design
- (a) outline schemes
- (b) development drawings
- (c) construction drawings
- (d) approval drawings
- (e) details
- (f) schedules
- (g) layouts

22.04.02 Surveys
- (a) topographical surveys
- (b) soil surveys
- (c) general surveys
- (d) pipeline surveys

22.04.03 Site development
- (a) earthworks
- (b) roads, paths and parking
- (c) railways
- (d) temporary works
- (e) general site works
- (f) fencing

22.04.04 Underground services
- (a) layouts
- (b) details
- (c) schedules
- (d) pipe anchors

22.04.05 Foundations and civils
- (a) piling layouts
- (b) foundations
- (c) bund and retaining walls
- (d) tank foundations
- (e) pipe tracks and anchors
- (f) paving
- (g) pits and basins
- (h) isolated pipe supports
- (i) bending schedules
- (j) holding-down bolt schedules
- (k) building foundations

22.04.06 Structural concrete
- (a) structures
- (b) slabs
- (c) blast walls
- (d) bridges
- (e) blast resistant buildings
- (f) bending schedules
- (g) details

22.04.07 Steelwork
- (a) structures
- (b) building frames
- (c) pre-assembled units
- (d) pre-assembled racks
- (e) pipe racks
- (f) platforms and ladders
- (g) pipe supports
- (h) schedules

22.04.08 Documentation
- (a) bills of quantity
- (b) specifications
- (c) contract documents

22.04.09 General activities

22.05 VESSEL AND HEAT TRANSFER ENGINEERING

22.05.01 Basic design: storage tanks
    (a) fixed roof tanks
    (b) floating roof tanks
    (c) double skinned tanks
    (d) spherical tanks
    (e) silos, bunkers and hoppers

22.05.02 Detailed design storage tanks
    (a) spiral stair and caged ladder
    (b) handrailing
    (c) roof walkway and interconnecting walkways
    (d) agitator fittings
    (e) coils internal and limpet
    (f) connections
    (g) support structure
    (h) double-skin cryogenic tanks

22.05.03 Basic design: pressure vessels and agitators
    (a) pressure vessels
    (b) separators
    (c) fermentors

22.05.04 Detailed design: pressure vessels and agitators
    (a) pressure vessel scantling calculations
    (b) supports
    (c) coils and jackets
    (d) agitator and supports
    (e) platform and clips
    (f) wind check and seismic loading
    (g) local loads
    (h) body flanges
    (i) internals

22.05.05 Basic and detailed design: columns and stacks
    (a) basic design calculations
    (b) body flanges
    (c) local loading
    (d) platforms, ladders and pipe clips
    (e) trays
    (f) packing
    (g) wind check and seismic loading
    (h) guy ropes

22.05.06 Basic and detailed design heat transfer
    (a) shell and tube exchangers
    (b) tank heaters
    (c) air coolers
    (d) cooling towers
    (e) waste heat recovery units
22.05.07 Fired heaters
    (a) boilers and incinerators
    (b) furnaces, ovens and kilns
    (c) flare systems
22.05.08 Disintegration equipment
    (a) crushers
    (b) grinders
    (c) mills
22.05.09 General activities

22.06 MACHINERY AND MECHANICAL ENGINEERING

22.06.01 Rotating machinery design
    (a) gas turbines
    (b) compressors: centrifugal
    (c) compressors: reciprocating
    (d) pumps: centrifugal
    (e) pumps: reciprocating
22.06.02 Mechanical handling design
    (a) conveyors
    (b) cranes and lifting equipment
    (c) unit loading systems
    (d) weighing equipment
    (e) stacking systems
    (f) storage systems
22.06.03 Mechanical packages design
    (a) water treatment
    (b) effluent treatment
    (c) packaged units
    (d) inert gas generators
    (e) steam-raising equipment
    (f) fuel systems
    (g) filtration units
22.06.04 Services
    (a) HVAC detail specification

            (b)   heating services
            (c)   air conditioning services
            (d)   plumbing
22.06.05   General activities

22.07   ELECTRICAL ENGINEERING

22.07.01   Specifications
            (a)   general electrical
            (b)   general motors
            (c)   mechanical package electrical
            (d)   equipment electrical
            (e)   installation
22.07.02   Requisition and data sheets
            (a)   main equipment requisitions
            (b)   bulk material requisitions
            (c)   material requisitions
            (d)   basic material and material lists
            (e)   calculation sheets
22.07.03   Diagrams
            (a)   main single-line diagram
            (b)   loading diagram
            (c)   logic sequence diagram
            (d)   control schematics
            (e)   lighting and small power schematics
            (f)   relay grading diagram
            (g)   wiring diagrams
22.07.04   Layouts
            (a)   power/substation
            (b)   communication
            (c)   cables
            (d)   cable tray
            (e)   earthing
            (f)   lighting and small power
            (g)   fire alarm
            (h)   trace heating
22.07.05   Schedules and schematics
            (a)   motor schedules
            (b)   terminal schedules
            (c)   cable schedules
            (d)   motor relay schedules
            (e)   cable hook-up diagrams

             (f)   communication diagrams
             (g)   fire alarm diagrams
22.07.06 Detail drawings
             (a)   earthing details
             (b)   cable installation details
             (c)   cathodic protection details
             (d)   trace heating isometrics
             (e)   hazardous area drawing
             (f)   civil requirement drawings
22.07.07 Manuals
             (a)   manuals and dossiers
             (b)   power study report
22.07.08 Safety
             (a)   safety guidelines
             (b)   safety review
             (c)   final safety review
22.07.09 General activities

22.08   INSTRUMENT ENGINEERING

22.08.01 Conceptual design
             (a)   philosophy diagrams
             (b)   logic diagrams
22.08.02 Layouts
             (a)   routes
             (b)   instruments
             (c)   control room
22.08.03 Hook-ups
             (a)   selection of hook-ups
             (b)   process allocation
             (c)   transmission allocation
             (d)   produce specials
             (e)   mounting and tracing
22.08.04 Indexes and schedules
             (a)   instrument index
             (b)   relief and safety device index
             (c)   input and output schedule
             (d)   equipment certification schedule
             (e)   cable list
             (f)   junction box schedule
22.08.05 Schematic diagrams
             (a)   loop diagrams

    (b) schematics
    (c) terminal indexes
    (d) field wiring block diagrams
    (e) power distribution diagrams and schedules
    (f) earthing diagrams
    (g) instrument utilities
22.08.06 Panel design
    (a) panel layouts
    (b) panel equipment lists
    (c) rack layouts
    (d) rack equipment lists
    (e) annunciator layouts
    (f) label lists
22.08.07 Instrument specifications
    (a) FM instruments (in-line)
    (b) FM instruments (field)
    (c) panel mounted instruments
    (d) reliability and calculations
22.08.08 System specifications
    (a) system specification
    (b) central control systems
    (c) programme logic control
    (d) panel and rack specifications
    (e) packaged equipment specifications
22.08.09 Material take-off
    (a) material specifications
    (b) material take-off
22.08.10 Process instrument specifications
    (a) control valves
    (b) relief valves
    (c) control instruments
    (d) indicating and alarm setting
22.08.11 Engineering line diagrams
    (a) preliminary
    (b) intermediate
    (c) final
22.08.12 General activities

22.09 PIPING ENGINEERING

22.09.01 Materials engineering
    (a) piping and valve specifications

	(b)	general piping specifications
	(c)	special piping design items
	(d)	flowsheet and line list review
22.09.02	Schedules	
	(a)	line
	(b)	design
	(c)	steam trap
	(d)	hose coupling
22.09.03	Piping models (notional)
	(a)	design model
	(b)	build base board
	(c)	model preliminary equipment
	(d)	model structure
	(e)	model cable and HVAC ducts
	(f)	design and model pipework
22.09.04	Piping models (detail)
	(a)	build base board
	(b)	model equipment
	(c)	model structures
	(d)	model cable and HVAC ducts
	(e)	model piping
	(f)	final ELD/ULD model check
22.09.05	Piping studies
	(a)	draft piping studies
	(b)	piping site survey
22.09.06	Layouts (manual method)
	(a)	plot plans
	(b)	equipment layouts
	(c)	escape route layouts
	(d)	HVAC layouts
	(e)	fire protection layouts
22.09.07	Layouts (computer method)
	(a)	build preliminary equipment
	(b)	build preliminary structure
	(c)	input preliminary cable and HVAC trays
	(d)	check plant model input
	(e)	equipment layouts
	(f)	escape layouts
	(g)	HVAC layouts
	(h)	fire protection layouts
	(i)	HVAC layouts
22.09.08	General arrangement drawings (manual method)
	(a)	general arrangement drawings

22.09.09 General arrangement drawings (computer method)
- (a) build approved equipment
- (b) build approved structure
- (c) input catalogues/specifications/instruments
- (d) input approved cable/HVAC trays
- (e) check plant model input
- (f) input pipes
- (g) interference connection checks
- (h) drawing production
- (i) input pipe supports
- (j) module administration

22.09.10 Isometrics (manual method)
- (a) prepare and complete isometrics
- (b) add welded attachments

22.09.11 Isometrics (computer method)
- (a) details validation
- (b) check isometric output

22.09.12 Piping stress engineering
- (a) stress piping
- (b) stress pipe supports
- (c) produce pipe stress sketches

22.09.13 Material take-off (manual method)
- (a) preliminary
- (b) intermediate
- (c) final

22.09.14 Material take-off (computer method)
- (a) specification input
- (b) materials control input
- (c) isometric production

22.09.15 Flow and line diagrams
- (a) process flowsheets
- (b) engineering line diagrams
- (c) utility line diagrams
- (d) HVAC flowsheets
- (e) fire protection flowsheets

22.09.16 General activities

22.10 SAFETY ENGINEERING

22.10.01 General safety considerations
- (a) procedures
- (b) bid evaluation

- 22.10.02 Safety criteria
  - (c) project control procedures
  - (d) safety supervision
- 22.10.02 Safety criteria
  - (a) design safety criteria
  - (b) design safety philosophy
- 22.10.03 Design safety
  - (a) fire and gas location drawings
  - (b) hazardous area drawing
  - (c) layout/passive protection/firewall location
  - (d) escape route drawing
  - (e) safety and fire equipment drawing
  - (f) safety philosophy
  - (g) HVAC damper operation
  - (h) safety signs drawing
  - (i) authority approval
- 22.10.04 Firewater distribution
  - (a) ring main calculations
  - (b) water demand calculations
  - (c) firewater requirement report
  - (d) firewater engineering line diagram
  - (e) area protection drawings
- 22.10.05 Safety and fire-fighting equipment
  - (a) deluge/sprinkler systems design
  - (b) deluge/sprinkler systems calculations
- 22.10.06 General activities

22.11 QUALITY ASSURANCE

- 22.11.01 Design phase
  - (a) project quality manual
  - (b) project quality plans
  - (c) project procedures
  - (d) design quality audits
  - (e) design reviews
  - (f) design document review
- 22.11.02 Procurement phase
  - (a) review requisitions
  - (b) review bid evaluations
  - (c) vendor quality audits
  - (d) review vendor quality assurance documents
  - (e) quality assurance/quality control clarification meetings

22.12 QUALITY CONTROL

22.12.01 Vendor quality control
- (a) QA/inspection: major purchases
- (b) QA/inspection: minor purchases
- (c) inspection co-ordination
- (d) review and approve Vendor QA documents

22.13 GENERAL ACTIVITIES

22.13.01 Procurement/expediting
22.13.02 Safety reviews
22.13.03 Approve client's drawings
22.13.04 Vendor print control and approval
22.13.05 Site liaison
22.13.06 Local authority liaison
22.13.07 Inter-departmental meetings
22.13.08 Site meetings and visits
22.13.09 Requisitions
22.13.10 Bid analysis sheets
22.13.11 As-built drawings
22.13.12 Construction support
22.13.13 Weight reports
22.13.14 Final documentation
22.13.15 Project close-out reports
22.13.16 Supervision

22.14 GENERAL PROJECT COSTS

22.14.01 Project office costs and expenses
- (a) overheads: head office
- (b) overheads: project office
- (c) computer time and equipment
- (d) publicity models
- (e) office space for client's personnel
- (f) overheads: photographic work and other publicity
- (g) special tests and development

22.14.02 Records and communications
- (a) preliminary documentation

        (b)   collation and storage of final documentation (inclusive of test and inspection records, operating and maintenance instructions and schedules)
- (c) special printing, duplicating and collating
- (d) photocopying
- (e) reproduction of drawings
- (f) microfilming and equipment
- (g) communications: telex, fax, couriers, etc.
- (h) special photography

22.14.03 Staff costs
- (a) travel expenses
- (b) subsistence
- (c) expenses for dependants, e.g. travel, accommodation
- (d) special payments bonuses
- (e) cost of client visits
- (f) personnel training
- (g) site entertainment (special visitors)

22.14.04 Fees and charges
- (a) consultants (various)
- (b) legal consultants
- (c) translation costs
- (d) royalties and licensing fees
- (e) licences and permits
- (f) insurance claims assessors
- (g) research and development
- (h) agent's commissions
- (i) inspection and testing authority fees, e.g. Lloyd's
- (j) insurance premiums
- (k) insurance of materials in storage
- (l) meteorological reports

22.14.05 Finance
- (a) escalation
- (b) professional insurance bonds
- (c) performance guarantees
- (d) finance costs and bank charges
- (e) forward buying of foreign currency
- (f) export credit guarantee premiums

22.14.06 Other requirements and services
- (a) pre-commissioning spares
- (b) commissioning spares
- (c) operational spares
- (d) overseas inspection services
- (e) vendor specialists and commissioning engineers

        (f)   outside buying agencies

22.14.07 Consignment costs
- (a) packing and protection
- (b) delivery to packing and forwarding agents
- (c) storage prior to shipment
- (d) dock dues and agent's fees
- (e) freight charges
- (f) import, export duties
- (g) bond area storage costs
- (h) freight insurance
- (i) storage at port of destination
- (j) freight from port to site

Notes